D0368714

# A PRIMER OF
# SOCIAL STATISTICS

# McGRAW-HILL SERIES IN
# SOCIOLOGY AND ANTHROPOLOGY

Richard T. LaPiere, *Consulting Editor*

*Baber*   MARRIAGE AND THE FAMILY

*Barnett*   INNOVATION: THE BASIS OF CULTURAL CHANGE

*Bergel*   URBAN SOCIOLOGY

*Bowman*   MARRIAGE FOR MODERNS

*Davie*   NEGROES IN AMERICAN SOCIETY

*Dornbusch and Schmid*   A PRIMER OF SOCIAL STATISTICS

*Gittler*   SOCIAL DYNAMICS

*Goode and Hatt*   METHODS IN SOCIAL RESEARCH

*Hoebel*   MAN IN THE PRIMITIVE WORLD

*Hoebel, Jennings, and Smith*   READINGS IN ANTHROPOLOGY

*House*   THE DEVELOPMENT OF SOCIOLOGY

*Landis*   RURAL LIFE IN PROCESS

*LaPiere*   A THEORY OF SOCIAL CONTROL

*LaPiere*   COLLECTIVE BEHAVIOR

*LaPiere*   SOCIOLOGY

*Lemert*   SOCIAL PATHOLOGY

*McCormick*   ELEMENTARY SOCIAL STATISTICS

*Queen and Carpenter*   THE AMERICAN CITY

*Schneider*   INDUSTRIAL SOCIOLOGY

*Tappan*   CONTEMPORARY CORRECTION

*Tappan*   JUVENILE DELINQUENCY

*Thompson*   POPULATION PROBLEMS

*Walter*   RACE AND CULTURE RELATIONS

*Young*   SOCIAL TREATMENT IN PROBATION AND DELINQUENCY

# A PRIMER OF
# SOCIAL STATISTICS

**SANFORD M. DORNBUSCH**

*Department of Social Relations*
*Harvard University*


**CALVIN F. SCHMID**

*Professor of Sociology*
*Director, Office of Population Research*
*University of Washington*

**1955**

McGRAW-HILL BOOK COMPANY, INC.
NEW YORK    TORONTO    LONDON

## A PRIMER OF SOCIAL STATISTICS

IV

THE MAPLE PRESS COMPANY, YORK, PA.

# Preface

The orientation of this primer reflects an attempt to overcome certain basic problems associated with the elementary course in social statistics. First, the majority of students registering for the beginning courses are so inadequately prepared in mathematics that only the most rudimentary materials can be presented. Students are justifiably fearful of their mathematical inadequacies and attempt to avoid the course. When they do take it, anxiety about their deficiencies frequently creates a defeatist attitude. From a long-term point of view, the most logical and constructive solution to this difficulty is, of course, additional training in mathematics both in high school and in college. Until this state of affairs is attained, however, the immediate problem must be faced realistically. This text attempts to bring statistics to the student instead of overwhelming him with the intricacies of more advanced statistical methods or indulging in vain exhortations that he gain literacy in mathematics. No knowledge of algebra is necessary for understanding the contents of this primer.

Second, students frequently are looking for a statistical cookbook. To them statistical analysis is a perfunctory mechanical process of applying formulas and operating a calculating machine. They merely want to grind out the right answers and pass the course. Without pandering to this tendency, all the more elementary computational procedures are presented in a graded and articulated manner. Emphasis, however, is placed on the nature of statistical reasoning, including appropriate explanations of the logic underlying various statistical concepts and

v

techniques. An attempt is made to show that statistics is a useful and necessary technique for solving problems and not merely a body of abstractions. In this connection, there are discussions on the derivation, application, and interpretation of various statistical tools.

Third, this primer obviously makes no pretense of surveying the entire field of statistics, nor does it represent a thorough explanation of the topics it does cover. Like most textbooks, it also cannot make any claim to originality with respect to content. Its claim to uniqueness lies more in its orientation, organization, and presentation.

The choice and presentation of topics are the outgrowth of considerable experience teaching elementary statistics, directing graduate theses, and supervising research organizations and projects. The subject matter and presentation follow a simple and gradual sequence of steps, each based on the material that precedes it. New concepts, principles, and processes are introduced with simple and concrete illustrative examples. Most of the illustrations and problems are based on actual research reported in books and periodicals, but the data frequently have been changed to simplify calculations. Emphasis is placed on those aspects of statistics which are most relevant and applicable to the social-science field, particularly sociology. An attempt has been made to cover essential statistical principles and practices without resorting to more technical and theoretical aspects of mathematics. When deemed appropriate, as well as compatible with the general level of presentation, mathematical derivations of formulas have been introduced. Emphasis is placed on the logic of statistical analysis rather than on computation skills. The simplest diction and writing style have been utilized. In short, every effort has been made to gauge the content, organization, and presentation of this primer in terms of the needs and abilities of the student.

Having mastered this book, the student should be able to understand reports of social research in which elementary statistical techniques are utilized. He cannot be expected to develop competence in undertaking his own research without additional training.

The authors will consider their endeavors a success if this primer helps the student to acquire (1) statistical literacy to the extent that he can recognize and understand basic statistical concepts, symbols, and techniques when he sees them; (2) an appreciation of statistics as a universal method of thinking, as well as a fundamental and essential tool for solving problems; (3) a critical awareness of the strength and weakness of statistical techniques; (4) a few basic elementary skills in statistical analysis.

The authors happily acknowledge their indebtedness to the following persons and organizations: to our colleagues, S. Frank Camilleri, Charles E. Bowerman, Richard J. Hill, and S. Frank Miyamoto, for providing us with an opportunity to use a preliminary version of this manuscript in their classes and obtain evaluations of its deficiencies; to Earle H. MacCannell for statistical assistance; to Eleonore Bristol, Donald P. Hayes, Donald C. Gibbons, William Eugene Torget, and Mildred Giblin for clerical and drafting assistance; to Julie M. Miyazaki for typing several drafts of the manuscript with exceptional competence; and to the following authors and publishers for permission to include material from their publications: Professor Sir Ronald A. Fisher, Cambridge, and Messrs. Oliver and Boyd, Ltd., Edinburgh; Professor George W. Snedecor, and the Iowa State College Press.

*Sanford M. Dornbusch*
*Calvin F. Schmid*

# Contents

# 1 Some Advice for the Student

The suggestions given in this chapter are designed to aid you in understanding, as well as remembering, the concepts, principles, and techniques presented in this book.

## 1-1 Statistics as a Language

Statistics has its own terms, words and symbols that are not used in the same way as in ordinary speech. In some respects, statistics is like a foreign language. In studying a foreign language, an entirely new vocabulary must be learned. Similarly, in statistics, the student will be required to learn the meaning of unfamiliar terms and symbols, but fortunately not as many as in a foreign language. For emphasis, as well as to facilitate study and review, each chapter is introduced with a listing of terms and symbols discussed in the chapter. You should be able to define each of the words and symbols listed after you have finished the chapter. If you are unable to do this, then further study is essential.

## 1-2 Mathematical Background

This text is designed for those with relatively little mathematical training or those who lack self-confidence in that area. If you are fairly proficient in mathematics and wish to cover certain subjects on a more advanced level, your instructor will be able to recommend supplementary materials.

In line with this nonmathematical emphasis, additional simplified explanations have been inserted in the text. When such explanations are included, the material previously covered is explained again in simpler terms. These inserts are set off in smaller type. If you understand the main presentation, the simplified explanations can be skipped.

## 1-3 Some Study Aids

The introduction at the beginning of each chapter attempts to explain briefly the purpose and content of the chapter. At the end of each chapter is a summary containing the more salient points covered. It is possible to memorize these summaries without reading the book. Obviously such a procedure is extremely shortsighted, since rote learning will not provide a base upon which later material can be added. In the end, it will give less knowledge and require more effort.

Many illustrative examples will be found throughout the book. Sometimes they are followed by problems to be done by the student. Correct answers always are included. It is strongly recommended that the student attempt to work out these problems. They are an invaluable aid in the learning process, as well as an index for determining strong and weak points.

In order to make study and review easier, each new central idea is numbered at the beginning of the paragraph in which it starts. The numbering system tells you first the chapter and then the discussion number. This explanation that you are reading is therefore part of the third section in the first chapter and accordingly is referred to as Sec. 1-3.

## 1-4 Daily Preparation

Finally, daily preparation is essential. This course, in general, is concerned with a different approach to the problem of making decisions. It takes time to assimilate this new outlook.

In addition, the material is cumulative. To understand the chapters on testing hypotheses, for example, you must apply almost all the material in the earlier chapters. Daily preparation for this course builds up a principal which will pay interest.

If you do your part, the result will be an experience which offers not only practical information and useful tools, but also an introduction to an exciting way to look at the world.

# 2   Tabular Presentation

**VOCABULARY**

continuous
discrete
array
class interval
frequency ($f$)

frequency distribution
tally form
true lower and upper limits
size of class interval ($i$)
mid-point of class ($m$)
open class
relative frequency distribution

In order to handle large bodies of data, possibly numbering in hundreds or even thousands of cases, it is necessary to devise simple and practicable methods of organizing them into some kind of concise and logical order. In tabulating large masses of data, some information is lost, but the gain in convenience is sufficiently great to offset the loss in precision and detail. This chapter explains the principles involved in organizing data in groups or classes in order to save time and effort.

## 2-1   Inaccuracy of Measurement

No measurement is ever absolutely exact. There is always some degree of approximation present, although it may be very slight, no matter how meticulous and skillful the observer or how precise his instruments or apparatus. When a man is said to be 71 inches tall, this means that he is closer to 71 than he is to 70 or 72. If a star is said to be 9,000 light-years away, this may imply that it is closer to 9,000 light-years than to 9,100. A student who receives 88 in a history examination is believed by the teacher to be closer to that grade than to 87 or 89. Even when the scientist strives for accuracy and describes the acceleration of a falling body as 31.9987 feet per second per second, there is still some degree of approximation. The acceleration, for example, may be somewhere between 31.9986 and 31.9987, but closer to the latter number.

## 2-2   False Accuracy

Sometimes the manner in which statistical data are presented gives a false impression of the degree of accuracy. The average grade received by a student on the same test taken three times in 10 minutes might be carried out to many decimal places when such a degree of accuracy is not actually present. If his grades were 66, 66, and 67, his average grade should not be given as 66.3333333. The mean cannot be more accurate than the numbers on which it is based, so 66 or 66.3 would be a more reasonable presentation in this instance. At certain military academies, grades are given to three decimal places in order to facilitate the scholastic ranking of the cadets. But a grade of 87.092 in calculus is not really more accurate than a grade of 87. Fictitious accuracy has been introduced in order to avoid ties. In social research, it is usually unwise to go beyond the accuracy of the original data.

## 2-3   Rounding

When rounding to the nearer of two numbers, results less than halfway between are given as the lower number. All numbers more than halfway are raised to the higher number.

**Illustration.** Round to the nearest hundred:

166; 434; 871; 1,432

*Ans.* 200; 400; 900; 1,400

Round to the nearest ten:

97; 114; 1,673; 8

*Ans.* 100; 110; 1,670; 10

Round to the nearest unit:

14.374; 86.4; 12.50001; 132.499

*Ans.* 14; 86; 13; 132

Round to one decimal place:

19.14; .16; 1,322.894

*Ans.* 19.1; .2; 1,322.9

Round to two decimal places:

83.466; .899; 36.14879

*Ans.* 83.47; .90; 36.15

Sometimes a number is exactly at the midway point, and it may be difficult to determine in which direction to round the number. For example, in rounding 34.5 it could be made either 34 or 35. In such instances it is the practice for some people to raise such numbers, of others to lower them. To follow either procedure is incorrect, since a bias may be introduced by always raising or always lowering a number. The rule in such situations, though arbitrary, must be one which, in the long run, will be equally likely to raise or lower a number. Such a rule is the one followed in this text. If a number is exactly halfway between two numbers, round to the even number.

**Problems. 1.** Round to the nearest unit:

18.5; 31.50; 11.501; 12.51; 187.5; 8.5

*Ans.* 18; 32; 12; 13; 188; 8

**2.** Round to one decimal place:

86.35; 144.49; 156.50001; 178.55

*Ans.* 86.4; 144.5; 156.5; 178.6

## 2-4   Continuous Variables

A *continuous* variable is one that can be theoretically measured to any desired degree of precision. The values of a continuous variable merge into one another by minute gradations. There is an unlimited number of possible values ranging between the lowest and the highest. Height and weight are examples of continuous variables. We say that one man weighs 162 pounds and another 163 pounds, but these statements are usually the result of rounding. Theoretically, if our instruments were sufficiently accurate, there is no value of weight that could not possibly be observed. One man could weigh 162.47893 pounds and another 194.934562987654 pounds. The degree of precision is purely the result of the accuracy of the measuring instrument.

## 2-5   Discrete Variables

Besides continuous variables there are *discrete* or discontinuous variables. Each value of a discrete variable is distinct and separate. That is, the values differ from each other by finite amounts. A man cannot have 3.45 children, but he can have either 3 or 4. The number of listeners to a television program might be 416,346, but not 187,345.89. Only certain integral results are possible. The number of Negroes in a community, the number of college students at a religious meeting, and the number of leaflets dropped from a plane are all examples of discrete variables.

## 2-6   Treatment of Discrete Variables as if They Were Continuous

Although an individual case or a single group is thought of as a discrete unit in statistical description, where many cases or groups are considered they can be treated legitimately as continuous variables. The average number of gainfully employed adults per family may be 1.94, although no such figure would be possible in a single family. The joke about 2½ children representing the typical family may be funny, but it does not invalidate the logic and usefulness of applying such a statistical concept.

Indeed, statisticians take even greater liberties with discrete variables. Often, they are treated as if they were continuous in order to simplify statistical manipulation. For example, if 64 persons are classified as migrants, the 64 may be treated as if values between 63.5 and 64.5 had been rounded to the nearest integer. The statistician knows that 64 in this case does not represent such an interval, but he finds it convenient to act as if this were so. This apparent laxity may have practical and worthwhile results, as will be observed later.

## 2-7  The Array

Table 2-1 presents the ages of 99 males arrested for forgery in Seattle, Washington, during the calendar years 1950 and 1951. Although there are only 99 cases, it appears as a confused jumble of figures. In order to discover and make clear the nature of the distribution of ages, the data have been organized into an *array* in Table 2-2. It will be observed that in the array the values (ages) have been arranged in order of magnitude. The items are arranged in ascending order, but it also would have been proper to use a descending order. Thus, the array provides a much

Table 2-1   Ages of Males Arrested for Forgery, Seattle, Washington: 1950 and 1951 (Original Data)

| Case | Age | Case | Age | Case | Age | Case | Age | Case | Age |
|---|---|---|---|---|---|---|---|---|---|
| 1 | 54 | 21 | 24 | 41 | 44 | 61 | 48 | 81 | 19 |
| 2 | 20 | 22 | 24 | 42 | 44 | 62 | 21 | 82 | 19 |
| 3 | 25 | 23 | 29 | 43 | 30 | 63 | 30 | 83 | 38 |
| 4 | 28 | 24 | 29 | 44 | 28 | 64 | 45 | 84 | 26 |
| 5 | 42 | 25 | 36 | 45 | 28 | 65 | 32 | 85 | 41 |
| 6 | 42 | 26 | 25 | 46 | 18 | 66 | 33 | 86 | 40 |
| 7 | 26 | 27 | 20 | 47 | 17 | 67 | 22 | 87 | 22 |
| 8 | 31 | 28 | 44 | 48 | 32 | 68 | 47 | 88 | 33 |
| 9 | 27 | 29 | 30 | 49 | 38 | 69 | 33 | 89 | 23 |
| 10 | 34 | 30 | 25 | 50 | 28 | 70 | 28 | 90 | 28 |
| 11 | 23 | 31 | 19 | 51 | 29 | 71 | 28 | 91 | 43 |
| 12 | 22 | 32 | 16 | 52 | 29 | 72 | 18 | 92 | 38 |
| 13 | 27 | 33 | 33 | 53 | 20 | 73 | 35 | 93 | 26 |
| 14 | 52 | 34 | 31 | 54 | 24 | 74 | 23 | 94 | 62 |
| 15 | 23 | 35 | 21 | 55 | 26 | 75 | 16 | 95 | 22 |
| 16 | 37 | 36 | 25 | 56 | 35 | 76 | 38 | 96 | 27 |
| 17 | 37 | 37 | 32 | 57 | 38 | 77 | 52 | 97 | 21 |
| 18 | 37 | 38 | 32 | 58 | 36 | 78 | 21 | 98 | 22 |
| 19 | 38 | 39 | 34 | 59 | 22 | 79 | 43 | 99 | 55 |
| 20 | 42 | 40 | 52 | 60 | 21 | 80 | 42 | | |

*Source:* Original records, Seattle Police Department.

clearer picture of the age pattern of a group of forgery suspects. The youngest arrestee for forgery is 16, and the oldest 62. Most of the forgers are in their 20s and 30s. The essential characteristics of a distribution can, however, be revealed even more effectively by grouping the data into classes. In fact, where hundreds or possibly thousands of cases are involved, such grouping becomes absolutely necessary.

## 2-8   Frequency Distributions

Let us set up groups for reorganizing the data in Table 2-2. Each group will be referred to as a *class interval*, the class being defined in terms of an interval. Each of the 99 cases will be assigned to the particular class within whose interval it happens

**Table 2-2   Ages of Males Arrested for Forgery, Seattle, Washington: 1950 and 1951** (Data Arranged in an Array)

| Case | Age | Case | Age | Case | Age | Case | Age | Case | Age |
|------|-----|------|-----|------|-----|------|-----|------|-----|
| 75 | 16 | 95 | 22 | 4 | 28 | 33 | 33 | 85 | 41 |
| 32 | 16 | 98 | 22 | 44 | 28 | 66 | 33 | 5 | 42 |
| 47 | 17 | 11 | 23 | 45 | 28 | 69 | 33 | 6 | 42 |
| 46 | 18 | 15 | 23 | 50 | 28 | 88 | 33 | 20 | 42 |
| 72 | 18 | 74 | 23 | 70 | 28 | 10 | 34 | 80 | 42 |
| 31 | 19 | 89 | 23 | 71 | 28 | 39 | 34 | 79 | 43 |
| 81 | 19 | 21 | 24 | 90 | 28 | 56 | 35 | 91 | 43 |
| 82 | 19 | 22 | 24 | 23 | 29 | 73 | 35 | 28 | 44 |
| 2 | 20 | 54 | 24 | 24 | 29 | 25 | 36 | 41 | 44 |
| 27 | 20 | 3 | 25 | 51 | 29 | 58 | 36 | 42 | 44 |
| 53 | 20 | 26 | 25 | 52 | 29 | 16 | 37 | 64 | 45 |
| 35 | 21 | 30 | 25 | 29 | 30 | 17 | 37 | 68 | 47 |
| 60 | 21 | 36 | 25 | 43 | 30 | 18 | 37 | 61 | 48 |
| 62 | 21 | 7 | 26 | 63 | 30 | 19 | 38 | 14 | 52 |
| 78 | 21 | 55 | 26 | 8 | 31 | 49 | 38 | 40 | 52 |
| 97 | 21 | 84 | 26 | 34 | 31 | 57 | 38 | 77 | 52 |
| 12 | 22 | 93 | 26 | 37 | 32 | 76 | 38 | 1 | 54 |
| 59 | 22 | 9 | 27 | 38 | 32 | 83 | 38 | 99 | 55 |
| 67 | 22 | 13 | 27 | 48 | 32 | 92 | 38 | 94 | 62 |
| 87 | 22 | 96 | 27 | 65 | 32 | 86 | 40 | | |

*Source:* Arranged from data in Table 2-1.

to fall. If, for example, there is a class interval labeled 20 to 24, this means that every arrestee for forgery whose age is 20, 21, 22, 23, or 24 will be placed in that class interval. When organized into class intervals, information about individual cases is lost. At the same time, however, a clearer understanding of the pattern of the group can be grasped.

It will be seen from Table 2-2 that there are 3 persons aged 20; 5, aged 21; 6, 22; 4, 23; and 3, 24. Every one of these cases will be assigned to the 20 to 24 class interval. The number of cases in each class interval is the *frequency* ($f$) of that class. The frequency of the 20 to 24 class is

$$3 + 5 + 6 + 4 + 3 = 21$$

Class intervals are set up so that they do not overlap. If a case could be placed in two or more class intervals, there would be no basis for deciding in which class it belongs. In addition, the class intervals must be inclusive, so that there is a class interval covering every observed case. We must be able to assign every case to one and only one class interval.

When all class intervals are properly arranged and the number of cases in each is counted, the result is a *frequency distribution*. Frequency distributions not only summarize and emphasize essential features of the data, but they also facilitate the computation of various kinds of statistical measures. These points will be covered in Chap. 8.

## 2-9  Tabulation Methods

The process of tabulating the frequency for each class of a frequency table can be routinized by the use of a *tally form*. Figure 2-1 shows a tally form, each slanted mark representing a case falling within the limits of the assigned class. It is a simple matter to count the marks and write the frequency for each of the classes.

In recent years the routine and time-consuming operations of tabulation as well as computation have been facilitated by the development of mechanic and electronic devices. The most widely used system records information on standard cards by

## AGE OF MALE SUICIDES
## SEATTLE: 1948-1952

| AGE | TALLY | FREQ. |
|-----|-------|-------|
| 15-19 | //// | 4 |
| 20-24 | /N/ /N/ // | 12 |
| 25-29 | /N/ /N/ /N/ /N/ /N/ // | 27 |
| 30-34 | /N/ /N/ /N/ /N/ /N/ /N/ //// | 34 |
| 35-39 | /N/ /N/ /N/ /N/ /N/ // | 27 |
| 40-44 | /N/ /N/ /N/ /N/ /N/ //// | 29 |
| 45-49 | /N/ /N/ /N/ /N/ /N/ /N/ /N/ | 35 |
| 50-54 | /N/ /N/ /N/ /N/ /N/ /N/ /// | 33 |
| 55-59 | /N/ /N/ /N/ /N/ /N/ /N/ /N/ / | 36 |
| 60-64 | /N/ /N/ /N/ /N/ /N/ /N/ /N/ //// | 39 |
| 65-69 | /N/ /N/ /N/ /N/ /N/ /N/ | 30 |
| 70-74 | /N/ /N/ /N/ /N/ /// | 23 |
| 75+ | /N/ /N/ /N/ /N/ //// | 24 |
| TOTAL NUMBER OF CASES | | 353 |

Fig. 2-1 Tally form for hand tabulation.

means of holes punched in coded positions on the cards. By means of electrical contact through the holes, tabulation and computation can be performed at a high rate of speed. The only manual operation required is the original punching of the data into the cards.

Figure 2-2 shows an IBM card reduced in size with data recorded for a single case of suicide. Immediately below the reduced card, the punched portion of the card is reproduced in full size. Every column has a particular meaning, and the 25 punched columns record considerable information. The first five columns show the case number, 00769. In this study row 1 in column 6 signifies a male, while row 2 in that same column indicates a female. Since row 1 in the illustration is punched,

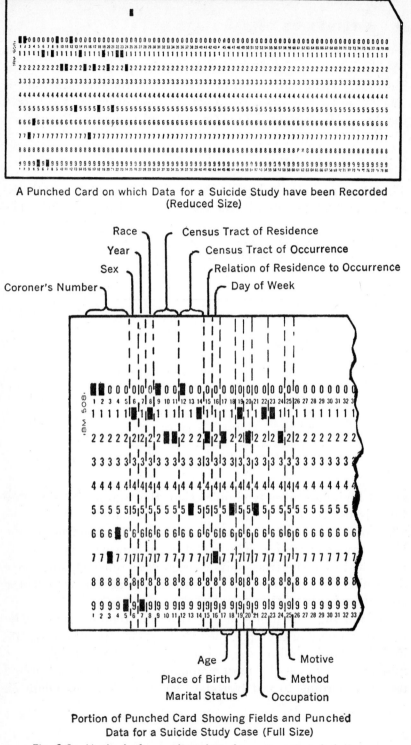

A Punched Card on which Data for a Suicide Study have been Recorded
(Reduced Size)

Portion of Punched Card Showing Fields and Punched
Data for a Suicide Study Case (Full Size)

Fig. 2-2 Method of recording data for automatic tabulation.

the case is a male. Column 7 records the year in which the suicide was committed, one of five years from 1948 to 1952. Since the code was based on the last digit of the year, this suicide was committed in 1949. In similar fashion other data collected for this study were coded and transferred to punched cards.

## 2-10   True Class Limits

The upper and lower class limits specified in a frequency distribution are not necessarily the *true lower* and *upper limits*. The

Table 2-3   Ages of Males Arrested for Forgery, Seattle, Washington: 1950 and 1951 (Data Arranged in a Frequency Distribution)

| Age | Frequency | Age | Frequency |
|-----|-----------|-----|-----------|
| Total | 99 | | |
| 15–19 | 8 | 40–44 | 11 |
| 20–24 | 21 | 45–49 | 3 |
| 25–29 | 22 | 50–54 | 4 |
| 30–34 | 15 | 55–59 | 1 |
| 35–39 | 13 | 60–64 | 1 |

*Source:* Compiled from data in Table 2-2.

stated limits are usually simpler than the true limits, since the table is designed to facilitate reading. Actually specifying the true limits of class intervals might be confusing, if not frightening, to the average reader. The statistician, however, is seriously concerned with the true limits of each class interval, since they determine the assignment of cases to classes, as well as provide a correct basis for further calculations.

To find the true limits in any situation requires knowledge, not technique. For example, the true limits for the class intervals in Table 2-3 depend on the method used by the police in determining ages. Since the most common procedure for recording ages is according to last birthday, let us evaluate our data on the basis of that assumption. If a man is 29 years plus 364 days old, his age is recorded as 29. The next day he would be 30.

On this basis, the lower limit of the class interval 15 to 19 is exactly 15. If a male is under 15 by even a day, he would not be placed in that class. The upper limit is not quite 20, and actually could be written as 19.99. The lower limit of the next higher class is 20; the upper limit is 24.99. The class interval 60 to 64 has the true limits 60 and 64.99.

Let us determine the true limits for the same class intervals when ages are recorded to the nearest year. The lowest class extends from 14.5 to 19.5, the next class from 19.5 to 24.5, etc. The upper limit of each class is the lower limit of the next class. There is overlapping at this single point. In actual practice, however, this is not a serious problem. Cases seldom have the precise values that would cause uncertainty in determining their class interval. If they do, they may be rounded to an even number (Sec. 2-3).

## 2-11  Size of Interval

The *size of the class interval* ($i$) is found easily. The lower limit of the class is subtracted from the upper limit, the result being the size of the class interval. Sometimes instead of referring to the "size" of a class interval, one may refer appropriately to the "width" or "length" of the class interval. Using the definition of age according to the last birthday, the size of the first class in Table 2-3 is 19.99 minus 15. The figure 19.99 is a short way of indicating that every value up to but not actually 20 is included in the interval. A more precise but cumbersome way of expressing this fact would be 19.9999999. For practical purposes, therefore, the size of the interval is 5. If age were designated to the nearest year, the size of the first class would be 19.5 minus 14.5, or 5.

## 2-12  Mid-point of Class Interval

For computations with grouped data, the *mid-point of the class* is used. The mid-point ($m$) is situated at half the distance between the true upper and lower limits. It can be found easily by adding together the upper and lower limits of the class and dividing by 2. In other words, the mid-point of the class is the average of the two limits.

The class 40 to 44 has a lower limit of 40 and an upper limit of 44.99 according to the first definition of age. The mean of 40 and 45 is 42.5, which is the mid-point of the class. If we use the second definition, the mid-point is $39.5 + 44.5$ divided by 2. The result is 42, showing that different concepts of age give different mid-points of class intervals.

Sometimes analysis of grouped data is based on the assumption that the mid-point of the class is equal to some kind of average of the values of the class. In performing computations it is assumed that all the cases have the value of the mid-point. In order to obtain reasonably accurate results, this assumption need not be exactly correct. If it were known that cases tended to concentrate at certain values, the proper procedure would be to set up the classes in such a way as to have the mid-points close to those values. For example, if, like Jack Benny, men often state their ages as 29 or 39, then the mid-points of such classes as 25 to 29 and 35 to 39 would deviate far from the average. It is a simple matter to select classes like 27 to 31 to take care of distributions with the type of concentrations indicated.

## 2-13   Open Classes

Suppose that Albuquerque Sam, aged 91, had been caught by the Seattle police after forging a check. In order to include him in the frequency distribution, additional classes would have to be added. If the same-sized intervals were used, six additional classes would be necessary: 65 to 69, 70 to 74, 75 to 79, 80 to 84, 85 to 89, and 90 to 94. The frequency of each of the first five classes would be 0 and the sixth class, 1. Instead of adding six new intervals to take care of one case, an alternative would be to add only one, labeled "65 and over." In terms of statistical practice this would be an acceptable substitute. A class which does not give the upper and lower limits, but instead is defined as more than or less than a specified limit, is called an *open class* or open-ended interval.

Open classes avoid the inclusion of superfluous classes, but unfortunately they have certain disadvantages. First, since only one limit is specified, there is no way of determining the mid-point of an open class. This may create considerable difficulty in computations with grouped data. Second, because of the indefiniteness of the open class, it is impossible for the reader to ascertain the approximate location of the cases within the class. For example, the man who is over 65 years of age might be 66, 86, 91, or 120.

## 2-14   Relative Frequency Distributions

Table 2-4 shows the frequency distribution of grades for two groups of students: those participating in extracurricular activities and those not so engaged. It gives considerable information, but because of the different number of persons in each group, direct comparisons between the groups are difficult.

Table 2-4   Frequency Distribution of Grade-point Averages for 150 Students Who Do Not, and 400 Students Who Do, Engage in Extracurricular Activities, Timbuktu Junior College, 1953

| Grade-point average | No extracurricular activities ($f$) | Extracurricular activities ($f$) |
|---|---|---|
| Total | 150 | 400 |
| .75– .99 | 10 | 50 |
| 1.00–1.24 | 30 | 30 |
| 1.25–1.49 | 50 | 80 |
| 1.50–1.74 | 20 | 80 |
| 1.75–1.99 | 10 | 50 |
| 2.00–2.24 | 5 | 40 |
| 2.25–2.49 | 10 | 30 |
| 2.50–2.74 | 15 | 30 |
| 2.75–2.99 | 0 | 10 |

Table 2-5   Relative Frequency Distribution of Grade-point Averages for 150 Students Who Do Not, and 400 Students Who Do, Engage in Extracurricular Activities, Timbuktu Junior College, 1953

| Grade-point average | Students with no extracurricular activities, per cent | Students with extracurricular activities, per cent |
|---|---|---|
| Total | 100.0 | 100.0 |
| .75– .99 | 6.7 | 12.5 |
| 1.00–1.24 | 20.0 | 7.5 |
| 1.25–1.49 | 33.3 | 20.0 |
| 1.50–1.74 | 13.3 | 20.0 |
| 1.75–1.99 | 6.7 | 12.5 |
| 2.00–2.24 | 3.3 | 10.0 |
| 2.25–2.49 | 6.7 | 7.5 |
| 2.50–2.74 | 10.0 | 7.5 |
| 2.75–2.99 | .0 | 2.5 |

Table 2-5 uses the same data in the form of a *relative frequency distribution* which gives the percentage of cases in each class rather than the frequency. Comparison is thereby made easy.

**Problems.** Using Tables 2-4 and 2-5, answer the following questions:

1. If a student in this group has a grade-point average over 2.50, is he more likely to be a participant in extracurricular activities or a nonparticipant?                    *Ans.* A participant

2. Of those students in extracurricular activities, what per cent have grades under 1.75?                    *Ans.* 60

3. Combining both groups of students, what per cent have grades under 1.50?                    *Ans.* 45.5

4. How many students who are not in extracurricular activities have averages over 2.00?                    *Ans.* 30

## 2-15   Determining Class Limits

Sometimes it may be found difficult to determine the most appropriate number of classes for a frequency distribution. If the classes are too numerous, some will contain relatively few cases or none at all. This makes the presentation more com-

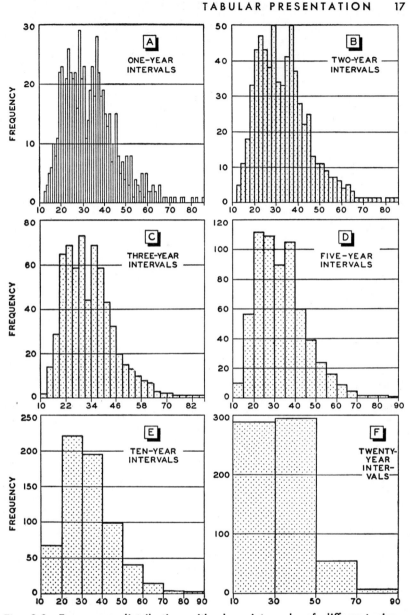

**Fig. 2-3  Frequency distribution with class intervals of different size.** Data represent the ages of 643 females who attempted suicide in the city of Seattle during the 5-year period 1948–1952.

Every table should be identified by a number

TABLE 1. MEAN RATES PER 100,000 OF POPULATION BY SEX AND AGE FOR COMPLETED SUICIDES, SEATTLE: 1948–1952

Every table should have a title

Captions, or column headings

Total may be placed either at top or bottom

Spacing and grouping of categories are important

Stubs, or row headings

Column figures should be properly aligned

| Age | Both sexes | Male | Female |
|---|---|---|---|
| Total | 20.1 | 30.2 | 9.8 |
| 10–14 | .9 | .... | 1.8 |
| 15–19 | 3.2 | 6.3 | .... |
| 20–24 | 9.8 | 12.9 | 6.6 |
| 25–29 | 18.7 | 25.7 | 11.6 |
| 30–34 | 24.3 | 35.8 | 12.7 |
| 35–39 | 27.4 | 29.7 | 25.1 |
| 40–44 | 26.0 | 34.1 | 17.8 |
| 45–49 | 28.9 | 44.4 | 13.0 |
| 50–54 | 32.4 | 46.1 | 19.1 |
| 55–59 | 33.7 | 54.5 | 13.3 |
| 60–64 | 36.9 | 62.6 | 10.1 |
| 65–69 | 33.6 | 59.2 | 7.9 |
| 70 and over | 36.5 | 75.0 | 4.0 |

Source of data should be indicated

*Source:* Calvin F. Schmid and Maurice D. Van Arsdol, "Attempted and Completed Suicides: A Comparative Analysis." *American Sociological Review*, **20** (June, 1955), in press.

Ruling at top and bottom and between columns

Explanatory footnotes also are placed beneath table

Fig. 2-4   Essential characteristics of a statistical table.

plicated, and many of the advantages of convenient summarization are lost. Also it is less likely that the mid-point of each class will be close to the mean of the class. On the other hand, if only a few classes are established, considerable information may be lost, and the frequency distribution fails of its primary purpose by concealing its characteristic structure.

There is no substitute for judgment in determining the most satisfactory number of classes in a frequency distribution. Generally the number of classes varies between 6 and 20, but there are exceptions even to this broad statement. The researcher must decide for himself what the optimal number is in each instance without allowing excessive detail or masking significant information. Figure 2-3 shows the result of using class intervals of different size. Both Fig. 2-3C and D bring out the essential characteristics of the distribution, whereas the other distributions are either too detailed or too general.

Calculations are made easier if all the class intervals are of equal size. Sometimes this may be awkward, and in such cases one must be very careful in finding the mid-point of each class. When classes are of unequal size, errors in calculation are more easily made.

Since the mid-points of class intervals will be used in later calculations, it is important that they should be convenient numbers in order to facilitate computation. Whole numbers, such as 4, 5, 10, 25, and 100, are relatively easy to work with. If cases tend to concentrate at certain values, an attempt should be made to make these values the mid-points of the class intervals, since it will be recalled the mid-point should be representative of the cases in each class (Sec. 2-12).

## 2-16  Essential Components of a Table

The essential parts of a table are illustrated in Fig. 2-4. The explanatory labels are designed to draw attention to the most noteworthy aspects of a statistical table.

## 2-17  Summary

A continuous variable is one that theoretically can attain any desired degree of precision; there is an imperceptible progression from one value to the next.

A discrete variable does not possess an infinite number of possible variations; rather the measures differ from each other by finite amounts.

Discrete variables are often treated as if they were continuous.

A frequency distribution shows the frequency of cases found in each class interval.

The size of a class interval is found by subtracting the true lower limit from the true upper limit.

The mid-point of a class interval is the average of the true upper and lower limits.

Computations with grouped data often assume that the mid-point of a class interval is the average of the cases found in that class.

It is impossible to find the mid-point of an open class.

A relative frequency distribution indicates the percentage of the cases in each class interval.

## PROBLEMS FOR CHAP. 2

**1.** Round to the nearest unit:

0.2; 37.5; 189.99; 40.5001; 34.5; 90.26; 53.55; 52.55

*Ans.* 0; 38; 190; 41; 34; 90; 54; 53

**2.** Arrange the following scores of a group of college freshmen on the Chapin social-insight scale in an array:

36; 45; 82; 23; 45; 57; 22; 43

*Ans.* 22; 23; 36; 43; 45; 45; 57; 82

**3.** Educational achievement is often recorded as the number of years of school completed. Accordingly, a person who has completed 8 years of grammar school and is now in his third year of high school would be credited with 10 years of school completed. The following table presents educational achievement for a group of industrial workers:

| Years of school completed | Frequency | Per cent of total |
|---|---|---|
| Total | 585 | 100 |
| 0–4 | 80 | 14 |
| 5–8 | 140 | 24 |
| 9–12 | 315 | 54 |
| 13 and over | 50 | 9 |

*a.* What is the sum of the percentages for the four classes?    *Ans.* 101

*b.* Why is the sum not exactly equal to 100?

*Ans.* Errors of rounding

*c.* How many workers finished 9 or more years of schooling?

*Ans.* 365

*d.* How many workers started to go to high school, whether they finished or not?    *Ans.* At least 365

*e.* What is the width of the second class interval, 5 to 8?

*Ans.* 4 years

*f.* What per cent of the workers finished a year or more of college?

*Ans.* 9

# 3 Graphic Presentation

**VOCABULARY**

rectilinear coordinates
$X$ axis
$Y$ axis
abscissa $(X)$
ordinate $(Y)$
frequency polygon
histogram
age and sex pyramid
smoothed frequency curve
positive skewness
negative skewness
J curve

U curve
cumulative frequency curve, or ogive
percentage change
semilogarithmic chart
bar chart
column chart
pie chart
statistical map
shaded map
spot map
isoline or isopleth map
maps with superimposed graphic forms
pictorial chart
three-dimensional charts

Statistical tables summarize masses of statistical data. Charts and graphs have a similar purpose. They are used to portray numerical facts in a simple and concrete manner. Since graphs and charts are essentially pictures of distributions, they are capable of arousing interest among readers, as well as stimulating analytical thinking.

## 3-1 Rectilinear Coordinates

Since a very large proportion of graphic techniques in the social sciences employ *rectilinear coordinates*, it is essential to understand their nature and applications.

The word "rectilinear" refers to the two lines or axes drawn at right angles (the lines are perpendicular to each other) by which points are located. First, two lines are drawn, intersecting at right angles. The horizontal line is called the *X axis*, and the vertical line is referred to as the *Y axis*. The *X* and *Y* axes are marked off into units. A distance on the *X* axis is called an *abscissa*, while a distance on the *Y* axis is labeled an *ordinate*. Units on both axes are used to locate points which represent values of an individual case with respect to both variables. If an attempt were made to study marital happiness in relation to income level, every couple would be assigned a score on each variable.

For example, the Wissenschaft family is assigned a score of 4 in marital happiness and 7 in income. If the marital-happiness score were labeled *X* and the income score *Y*, the Wissenschaft family would be symbolized by (4,7). The abscissa always is written first; the ordinate, second. This rule should be rigorously followed; otherwise it would be difficult to distinguish the Wissenschaft scores from the Pizza scores, whose marital-happiness rating is 7 and income 4. The Pizza coordinates would be written as (7,4), since the *X* value is written first.

Now let us plot the values for each family as a single point on a rectangular grid. For the Wissenschafts, the abscissa is 4, so it is necessary to move along the *X* axis up to the point labeled 4. But the *Y* value of 7 must be taken into account simultaneously. Every point directly above 4 has the same *X* value, but the point that is on a line with 7 on the *Y* axis is the only one that satisfies both conditions (4,7). The same point would be reached by first finding 7 on the *Y* axis, imagining a horizontal line at height 7, and then finding the point on that line that is directly above 4. Figure 3-1 shows the location of scores for four different families, including the Pizzas and Wissenschafts.

In accordance with the preceding explanation, the data of a frequency distribution can easily be portrayed graphically. The *X* axis, or horizontal scale, is divided in accordance with the class intervals in the frequency table. The *Y* axis is constructed to portray frequencies. The *Y* axis must always begin with zero and be devised so as to accommodate the maximum

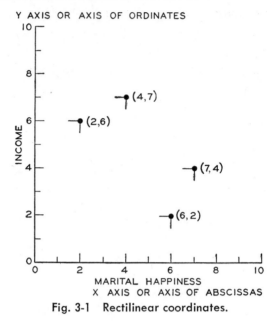

Fig. 3-1   Rectilinear coordinates.

class frequency. Scale divisions for the vertical axis are usually expressed in convenient numbers, such as units of 5 or 10 or 100.

## 3-2   Frequency Polygon

In a *frequency polygon*, points are plotted to represent the frequency of each class, and these points are then joined by straight lines. The frequencies represent the number of cases in the whole interval, yet the points represent only one $X$ value. Obviously, the point chosen as the abscissa must stand for the whole class. From this point of view, the mid-point of each class is the abscissa for each interval in a frequency polygon. Each point represents the frequency $(Y)$ of that class whose mid-point is $(X)$. To illustrate the essential characteristics of a frequency polygon, the data from Table 2-3 are portrayed in Fig. 3-2.

Two or more frequency distributions can be compared graphically by means of frequency polygons. When, however, the total number of cases in the two distributions is very different, graphic comparison of frequencies may be difficult to interpret. Accord-

DISTRIBUTION OF MALES ARRESTED FOR FORGERY
SEATTLE: 1950-51

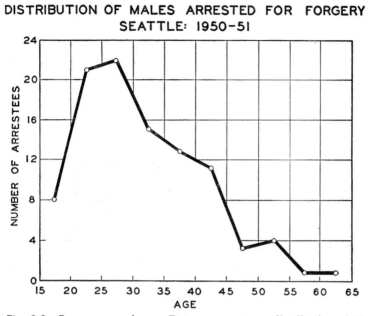

Fig. 3-2  **Frequency polygon.** Data represent age distribution of 99 males arrested for forgery in the city of Seattle in 1950 and 1951.

ingly, it is recommended that the numerical frequencies be transposed into percentages and the percentages plotted on the $Y$ axis.

## 3-3  Histogram

A second method of portraying frequency distributions is the *histogram.* For the histogram the $X$ and $Y$ axes are constructed in exactly the same manner as for the frequency polygon. The only difference is that the histogram utilizes areas of rectangles in representing class frequencies. These rectangles are constructed by erecting vertical lines at the upper and lower limits of each class interval (Fig. 3-3).

The height of each rectangle in a histogram is influenced not only by the frequency of the class but by the width of the class interval. The area of the rectangle represents the number of cases. The area is found by multiplying the base times the

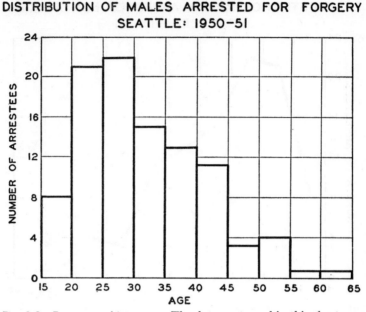

Fig. 3-3  Frequency histogram. The data portrayed in this chart are
the same as those in Fig. 3-2.

height. If two classes have the same size interval, then the
height will directly express the frequencies of the two classes.
But if one class is twice as wide as another, then identical
frequencies will be expressed by having the height of the wider
rectangle one-half the height of the second. This is not as
unreasonable as it seems. The wider class contains more scores,
and having the same frequency indicates that these scores are
not as likely to occur as the scores in the narrower class interval.
It is important to remember that the frequency of a class is
represented in a histogram by the area of a rectangle.

A variation of the simple histogram is the *age and sex pyramid*.
It is frequently used in ecological and population studies. The
histogram rectangles to the left of the zero line represent the
male sex, and those to the right the female sex. Actually the
age and sex pyramid is a double histogram with the $X$ and
$Y$ axes in reverse positions. Figure 3-4 is an illustration of this

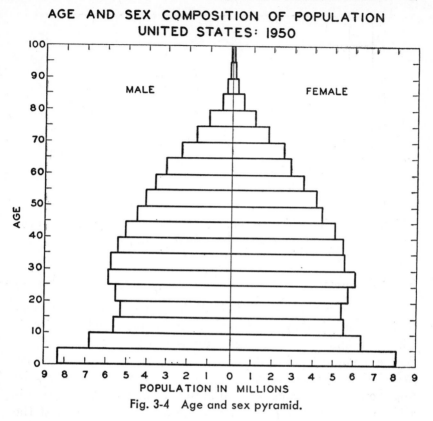

## AGE AND SEX COMPOSITION OF POPULATION UNITED STATES: 1950

Fig. 3-4 Age and sex pyramid.

type of chart. The horizontal axis represents frequencies, and the vertical axis age.

## 3-4 Smoothed Frequency Curve

Sometimes a frequency polygon or histogram is transformed into a *smoothed frequency curve.* The smoothing process attempts to eliminate irregularities considered accidental or the result of chance processes. The smoothing can be done freehand, by mechanical means, or by mathematical calculation. Essentially, the smoothing of a frequency curve represents a decision by the researcher to eliminate chance variability in his sample results and present a general characterization of the universe from

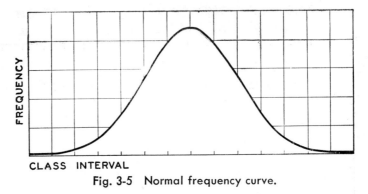

CLASS INTERVAL

Fig. 3-5   Normal frequency curve.

which he obtained his sample. Figure 3-5 shows a normal curve as one example of a smoothed frequency curve.

## 3-5   Skewness

It must not be assumed that all frequency curves are symmetrical and single-peaked as in Fig. 3-5. In actual practice, it will be found that frequency curves are to a greater or less degree asymmetrical or skewed. *Skewness*, a lack of symmetry, may be *positive* or *negative*. If there are extreme cases at the upper end of the scale, the tail is to the right, and the curve is said to be skewed positively. If there are extreme cases at the lower values, the tail of the frequency curve is to the left and is said to be skewed negatively. Figure 3-6 illustrates positive and negative skewness. Skewness will be discussed further in Chap. 13.

## 3-6   Special Curves

Certain types of frequency curves with unusual shapes have been given special names. These names are descriptive of the appearance of the curves. The *J curve*, which looks like the letter J, or a reverse J, is one of these. Frequency curves of this type often occur in studies of conforming behavior. Most people conform to the social norm, and the greater the deviation from the norm, the fewer the number of deviants.

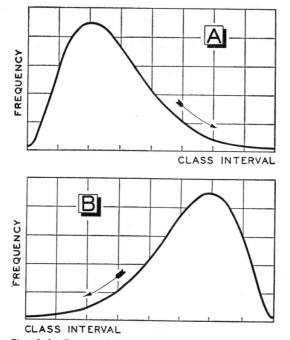

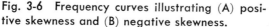

Fig. 3-6  Frequency curves illustrating (A) positive skewness and (B) negative skewness.

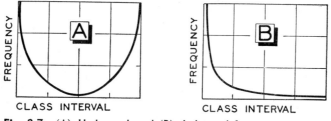

Fig. 3-7  (A) U-shaped and (B) J-shaped frequency curves.

Another type of curve is the *U curve*, which, as might be inferred, resembles the letter U. Both the J-shaped and U-shaped frequency curves are illustrated in Fig. 3-7.

## 3-7  Ogives

The *cumulative frequency curve*, or *ogive*, is used to portray graphically cumulative frequency distributions. The accumu-

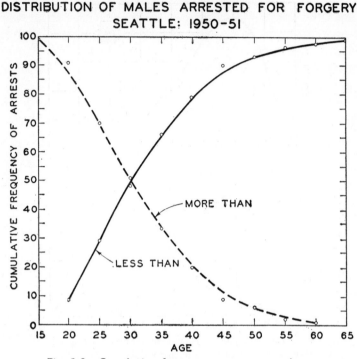

DISTRIBUTION OF MALES ARRESTED FOR FORGERY
SEATTLE: 1950-51

Fig. 3-8    Cumulative frequency curves, or ogives.

lated frequencies are represented by the vertical axis, and the
class intervals by the horizontal axis. The high point on the $Y$
axis indicates the total number of cases in the distribution.

There are two types of ogives: *more than* and *less than*. In
order to show how many cases are above a specified point or
value, a "more than" ogive is plotted. The "less than" ogive
indicates how many cases are below a particular value. The
two types of ogives are easily distinguishable, since the more-
than curve will have its maximum cumulative frequency at the
lower end of the $X$ scale, while the less-than ogive will have its
maximum cumulative frequency at the higher end of the $X$ scale.
Cumulating all the frequencies above each interval gives a more-
than ogive.

Instead of plotting the cumulated frequencies at the mid-
point of each class, as was done for the frequency polygon, they
are plotted at the upper and lower limits of each interval. For

the less-than ogive the cumulative frequencies are plotted at the upper limits of the class intervals, and for the more-than ogive at the lower limits of the class intervals. The reason for following this procedure is quite simple. When working with grouped data, one knows only the number of cases above or below a point when that point is at the boundary of a class. The way the cases are distributed within the interval is unknown, although certain reasonable assumptions can be made. Figure 3-8 represents both the less-than and more-than ogives based on the data in Table 2-3. As is usually the case, ogives take on the characteristic shape of elongated S's.

## 3-8   Time Series

So far most of this chapter has been concerned with a special kind of rectilinear coordinate graph in which the ordinates

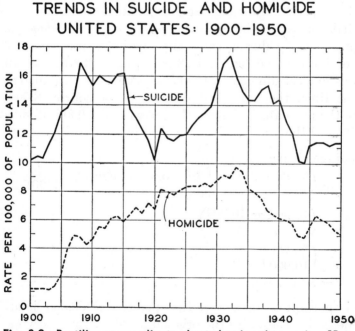

TRENDS IN SUICIDE AND HOMICIDE
UNITED STATES: 1900–1950

**Fig. 3-9   Rectilinear coordinate chart showing time series.** Note that the horizontal, or X, axis represents time, and the vertical, or Y, axis rates per 100,000 of population. Since the basic data represent annual rates, there are 51 plotting points for each series.

express some kind of frequency. Another type of rectangular grid is used for *time series*. The ordinal values can be expressed in any kind of numerical units, such as index numbers, percentages, or frequencies, but the abscissal values are used to represent a time dimension. The horizontal axis may indicate hours, days, weeks, months, or years as its units, since any measure of time can be used for the abscissal values in a time series. The graphic portrayal of a time series makes it easy to compare variations and trends during some period. Figure 3-9 illustrates how a rectangular coordinate chart may be used to portray changes in suicide and homicide rates over a 50-year period.

## 3-9   Percentage Change

The presentation of a time series on a rectangular grid enables the reader to observe the amount of increase or decrease during any time period. Rectangular coordinate graphs are not well adapted for portraying percentage change. *Percentage change* indicates change during some time period expressed as a per cent of the original figure. If, for example, there were 10,000 Italians in a certain city in 1940 as compared with 18,000 at the present time, the numerical increase is, of course, 8,000. To derive the percentage change, the increase of 8,000 is divided by 10,000, and the quotient is multiplied by 100.

Percentage change

$$= \left( \frac{\text{figure for later date} - \text{original figure}}{\text{original figure}} \right) (100)$$

$$= \left( \frac{18,000 - 10,000}{10,000} \right) (100)$$

$$= \left( \frac{8,000}{10,000} \right) (100)$$

$$= 80 \text{ per cent}$$

It is important to remember that percentage change is always considered in relation to the original number. By observing this convention, one never gets results like "a decrease of 150 per cent." No number can decrease more than 100 per cent, because at that point we have reached zero. If the Italian population declined from 10,000 to 1,000, the percentage change would be −90 per cent. It is not down 900 per cent, the result obtained by using the final figure as the base.

## 3-10   Semilogarithmic Chart

The *semilogarithmic chart* is unsurpassed for portraying percentage change. In addition to correctly representing relative changes, it also indicates the amount of increase or decrease at the same time. The characteristic structure of the semilogarithmic chart is a horizontal axis with arithmetic rulings, usually in time units, and a vertical axis with logarithmic rulings. The continued narrowing of spacings of the scale divisions on the vertical axis is characteristic of logarithmic ruling, while equal intervals on the horizontal axis indicate arithmetic ruling.

**Supplementary Explanation.** The logarithm of a number is the power of 10 which will give that number.

| | |
|---|---|
| The logarithm of 10 is 1 | $(10) = 10 = 10^1$ |
| The logarithm of 100 is 2 | $(10)(10) = 100 = 10^2$ |
| The logarithm of 1,000 is 3 | $(10)(10)(10) = 1,000 = 10^3$ |
| The logarithm of 10,000 is 4 | $(10)(10)(10)(10) = 10,000 = 10^4$ |

The logarithm of 1 is 0. This can be readily explained. Any number to the power of 0 will give a result of 1, so that $10^0$ is equal to 1. The rule that the zero power of a number equals 1 is necessary to fit into the rules for multiplication and division of numbers. When dividing $10^4$ by $10^3$, we merely subtract the exponents, giving $10^{4-3}$, or $10^1$. The same procedure for dividing $10^4$ by $10^4$ gives a result $10^{4-4}$, or $10^0$. But it is obvious that dividing a number by itself, as in the case of $10^4$ by $10^4$, must give an answer of 1. To preserve consistency, therefore, it is necessary that $10^0$ equal 1.

The logarithm of every number from 1 to 10 is between 0 and 1. The logarithm of every number from 10 to 100 is between 1 and 2. In a table of common logarithms one can find the log of any number by remembering between which two numbers (two powers of 10) the logarithm must lie.

If the population of one community increased from 100 to 200 and the population of another city increased from 500 to 1,000, the second city would have a greater amount of increase. But in statistical work, relative or percentage changes are sometimes more significant than absolute changes, changes in amount. The percentage changes in the foregoing example are identical, both communities showing increases of 100 per cent. Let us note how the logarithms of the four numbers will indicate an equal percentage change.

If we had a table of common logarithms (omitted from this text because this is the only chapter in which logarithms are discussed), we would discover that the logarithms of 100, 200, 500, and 1,000 are as follows:

$$\log 200 = 2.30103 \qquad \log 1{,}000 = 3.00000$$
$$\log 100 = 2.00000 \qquad \log 500 \;\; = 2.69897$$

A logarithmic scale will correctly show the comparative percentage increase in each community. Since the percentage increases are equal, the increases in logarithms for the two communities should be equal. Subtracting the log of each original figure from the log of each final population gives .30103 in each case. The equal change in the logarithms indicates equal percentage increases. Such relative changes are portrayed without distortion on a semilogarithmic chart, whereas a rectilinear coordinate chart shows only the amount of increase or decrease, not the percentage change.

If a semilogarithmic chart is examined carefully, it will be observed that:

*a.* There is no zero point on the logarithmic scale, nor are there any negative numbers. This occurs because any power of 10, negative or positive, will give a positive answer.

*b.* The values at the beginning of each cycle of numbers on the logarithmic axis are some multiple or division of 10.

*c.* The distances between the values in each cycle are unequal, becoming progressively smaller as the values increase.

*d.* If pairs of numbers are in the same ratio to each other, the distances on the logarithmic scale must be equal. For example, the distance from 100 to 500 is equal to the distance between 1,000 and 5,000.

*e.* The slope of the curves for any given period is indicative of

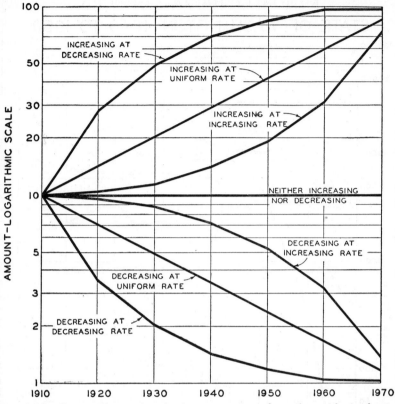

**Fig. 3-10**   Curves illustrating the interpretation of semilogarithmic charts.

the rate of change of the variable. If the slope of the line is sharp, then the rate of change is relatively great; and, by the same logic, a gradual rate of change is indicated by a comparatively slight slope. It makes no difference on what part of a semilogarithmic chart a curve is located; the same slope means the same rate of change.

Figure 3-10 depicts a series of type curves on a semilogarithmic grid. The following characteristic curve configurations are illustrated: (a) increasing at a decreasing rate, (b) increasing at an increasing rate, (c) increasing at a constant rate, (d) decreasing at a decreasing rate, (e) decreasing at an increasing rate, (f) decreasing at a constant rate, and (g) neither increasing nor decreasing. Figure 3-11 based on actual data illustrates further

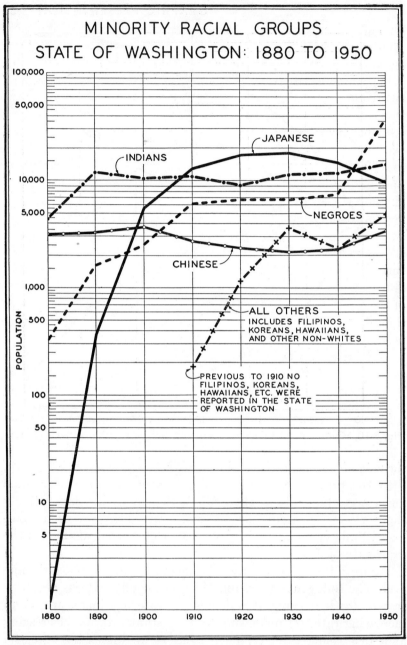

**Fig. 3-11 Semilogarithmic chart.** In this type of chart the relative slope of a curve indicates rate of change. Note that there are five decks or cycles in this chart, and the values portrayed range from 2 to over 30,000.

the outstanding characteristics of the semilogarithmic chart. This chart portrays almost all the basic patterns that will be found in Fig. 3-10.

Perhaps the essential features and application of the semilogarithmic chart can be revealed best by plotting the same data on both arithmetic and semilogarithmic grids that have been placed in juxtaposition (Fig. 3-12). The arithmetic chart may conceal and distort facts. One receives the impression that the rate of population growth for Dallas is very much less than that for the state of Texas as a whole. Moreover, it is virtually impossible to interpret even approximately what the population of Dallas was during the early period. This condition is inevitable if the range on the vertical scale of an arithmetic chart is very wide. On a logarithmic scale this makes no difference; it is possible to compare a large number of curves of widely different values on a semilogarithmic chart with clarity and precision. Moreover, the relative changes, as indicated by the slopes of the curves, are comparable no matter what their position may be on the grid. It will be seen from Fig. 3-12 that

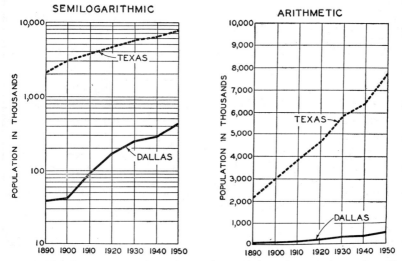

Fig. 3-12  Comparison of semilogarithmic and arithmetic charts. It will be seen that the semilogarithmic chart is superior to the arithmetic chart in portraying rate of change.

Dallas has manifested a much more rapid rate of growth than the entire state of Texas during this 60-year period.

## 3-11    Standards of Graph Construction

Although the primary purpose of this chapter is to help develop ability to read and understand various types of statistical charts, a few brief remarks on standards of graph construction may be pertinent. For anyone seriously interested in social research, a knowledge of the principles of constructing graphs and charts is indispensable. The average research worker may never actually draw many charts, but he most certainly will have to plan them and direct draftsmen in executing them in their final form. The following principles are particularly applicable to rectilinear coordinate charts:

*a.* Every chart should have a title which is usually placed at the top. As a working principle the title should answer three basic questions: What? Where? When?

*b.* The scales should be carefully laid out so there are not too many or too few divisions. There always should be a zero or some other base line for the ordinal axis. If only the upper part of the grid is needed, the scale can be broken, but at the same time the base line should be retained.

*c.* There should be scale figures for both axes with, of course, appropriate scale legends.

*d.* Grid lines should be drawn relatively lightly and kept to a minimum. No definite rule can be specified as to the optimum number of lines, since the size and other features of the chart must be taken into consideration.

*e.* It is of the utmost importance that the scale proportions be kept in proper balance. If, for example, the vertical axis is drawn too high in relation to the horizontal axis, the curve movements will tend to be exaggerated.

*f.* The curves on the grid should be drawn as heavy lines, and in case there is more than one they should be differentiated by distinct, carefully drawn patterns.

*g.* In order to facilitate interpretation, each curve should be labeled clearly and unmistakably.

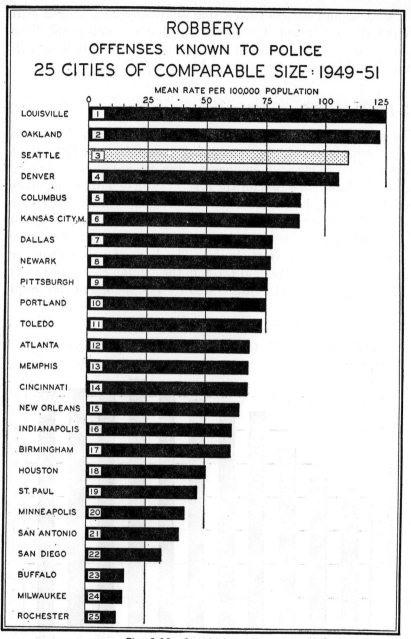

ROBBERY

OFFENSES KNOWN TO POLICE

25 CITIES OF COMPARABLE SIZE: 1949-51

MEAN RATE PER 100,000 POPULATION

Fig. 3-13  Simple bar chart.

*h.* It should be recognized that too many curves on a chart may cause confusion. The size of the chart, as well as the distribution of the curves in the field, would naturally influence the number that might be portrayed properly.

### 3-12    Bar and Column Charts

In graphic presentation, the simplest and most exact comparisons can be made by comparing length, a one-dimensional comparison. This fact emphasizes the advantage of using *bar charts* and *column charts* for the presentation of data. Comparison of the relative sizes of areal (two-dimensional) and cubic (three-dimensional) forms is more difficult. The length of each bar or column is proportional to the value portrayed. Bar charts and column charts are very similar. The bars of a bar chart are arranged horizontally, and those in the column chart, vertically. The simple bar chart is illustrated by Fig. 3-13, and the simple

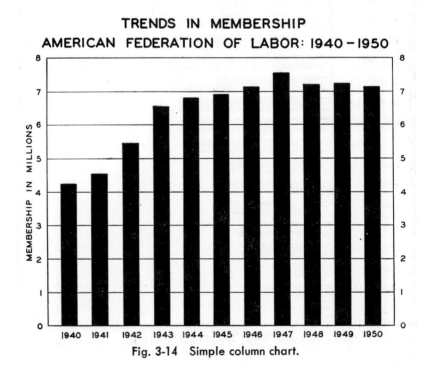

Fig. 3-14   Simple column chart.

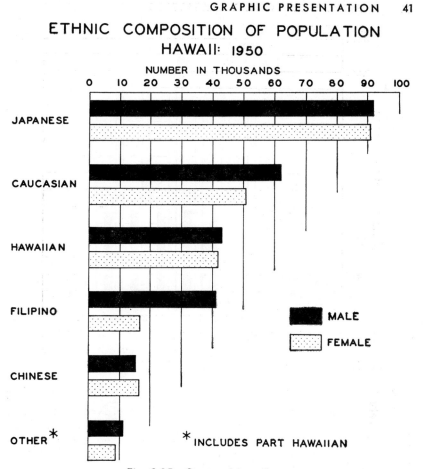

# ETHNIC COMPOSITION OF POPULATION
## HAWAII: 1950

NUMBER IN THOUSANDS

Fig. 3-15  Grouped bar chart.

column chart by Fig. 3-14. There is one bar or column for each category in simple bar or column charts, while grouped bar and column charts, as in Figs. 3-15 and 3-16, have two or more bars or columns for each category.

In constructing bar and column charts, careful consideration should be given to the following standards and principles of design:

*a.* The bars should be arranged in some systematic order, usually according to magnitude, starting with the largest. In a column chart, where a time series is portrayed, the columns, of course, should be in chronological order.

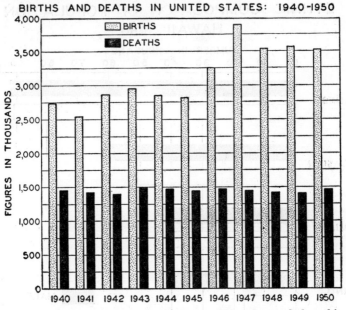

**Fig. 3-16  Grouped column chart.** It will be observed that this chart depicts the actual number of births and deaths in the United States for each year from 1940 to 1950.

*b.* The width of the bars and columns, as well as the spacing between them, possesses no special significance. Needless to say, for any particular chart the bars and columns should be of uniform width and properly adapted to the over-all size, proportion, and other features of the chart.

*c.* As a general practice a scale should be used in every bar or column chart. The number of intervals on the scale should be adequate for measuring distances but not too numerous to cause confusion. The intervals should be indicated in round numbers, perferably in such units as 5s, 10s, 25s, 50s, and 100s. The scale of bar and column charts should begin always with zero, and never be broken.

### 3-13  Pie Chart

Although the *pie chart* enjoys much popularity and is widely used in newspapers and other mass media of communication, it has

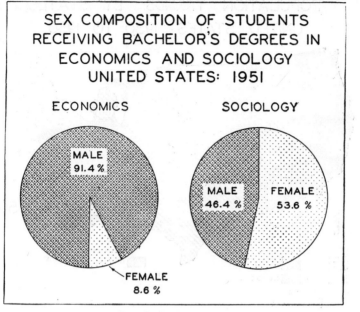

Fig. 3-17    Pie chart.

certain inherent weaknesses. First, the pie chart may become too complicated for ready and reliable visual comparison, particularly if too many divisions are shown. Almost never can more than five components be shown on a pie chart. Second, the pie chart is perceived generally as an areal comparison, which is more difficult than linear comparison. The pie chart is used to portray proportions of a total. The size of each slice (sector) is proportional to the value of each component. Figure 3-17 shows the proportion of men and women receiving college degrees in the United States in 1951 in economics and sociology. Each pie is divided into two sectors, one showing the percentage of men and the other of women. The total area of the pie represents 100 per cent.

## 3-14   Statistical Maps

Ecological (spatial) data are often used in the social sciences, particularly in population studies or in human ecology. Such

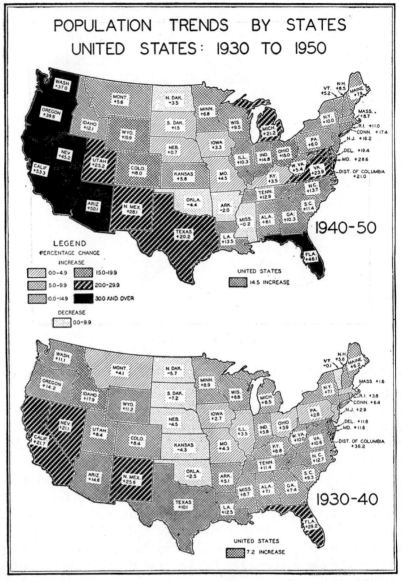

**Fig. 3-18** **Crosshatching technique showing rates of population change for two different periods.** It will be observed that the hatching gradation is related to rates of change.

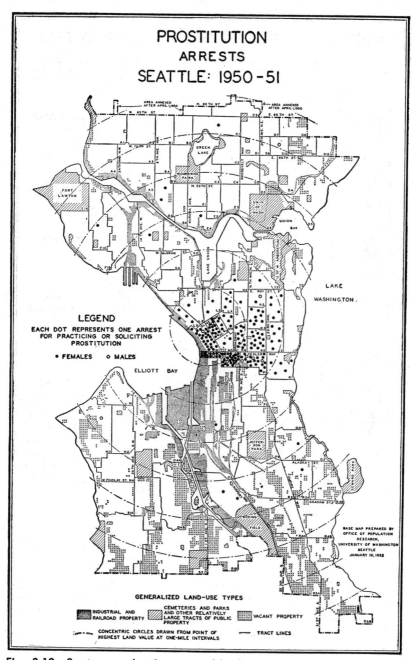

**Fig. 3-19 Spot map showing countable frequencies.** Male and female arrestees for practicing or soliciting prostitution are differentiated by special symbols. Note base map with generalized land use, census tracts, and concentric circles.

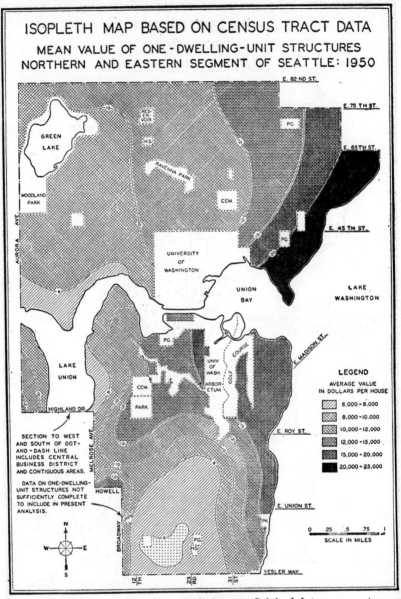

**Fig. 3-20 An illustration of an isopleth map.** Original data represent mean values of one-dwelling-unit structures for 36 census tracts.

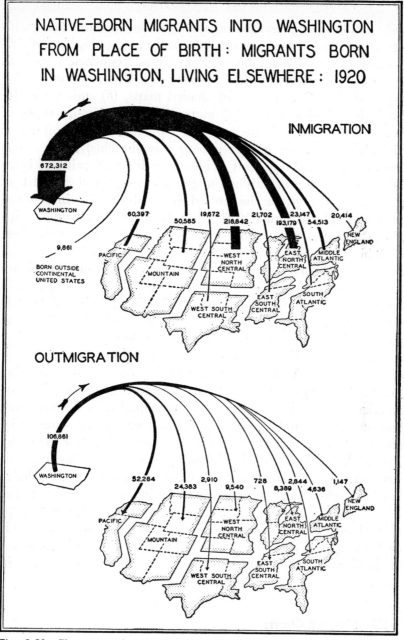

**Fig. 3-21 Flow map showing population migration.** The volume and direction of inmigration and outmigration to Washington state are clearly shown by this technique. The width of the lines indicates the number of migrants.

data can best be presented on *statistical maps*. Rates and ratios are often plotted on base maps to indicate geographical distributions of social phenomena. Sometimes frequencies, such as total population, are plotted on base maps. There are four major types of statistical maps: (*a*) shaded maps, (*b*) spot maps, (*c*) isopleth or isoline maps, (*d*) maps with one or more types of graphs superimposed, such as the bar, column, line, flow, or pictorial forms. On a *shaded map* (crosshatched), gradations from light to dark are used to illustrate different values. Figure 3-18 shows two crosshatched maps. The maps portray rates of population change for the United States during the intercensal periods 1930–1940 and 1940–1950. The lightest shading (stippling) indicates population decreases, and the darker shadings (hatching) population increases. *Spot maps* use spots or symbols to indicate the more or less exact location of phenomena. Figure 3-19 illustrates a simple spot map. The data represent arrests for practicing or soliciting prostitution in the city of Seattle during the 2-year period 1950–1951. The black circles indicate females, and the open circles males. The size, number, density, shading, or form of the spots can be used to present certain types of spatial data in visual form. The *isopleth* ("isos," meaning "equal" and "pleth," meaning measure) or *isoline map* is characterized by a series of lines connecting equal values. Along each line one would find an equal frequency or rate of some social phenomena. An illustration of an isoline map is portrayed by Fig. 3-20. The isolines indicate mean values of one-dwelling-unit structures in the northern and eastern segment of Seattle. A hatching scheme has been applied to bring out more clearly the isoline-value gradations. An example of the fourth major type of statistical map in which various graphic forms may be used is depicted by Fig. 3-21. This chart is a *flow map* indicating streams of inmigration and outmigration for Washington state in 1920.

## 3-15    Pictorial Charts

In recent years there has been an increasing use of pictorial symbols to represent statistical data. The use of such symbols is

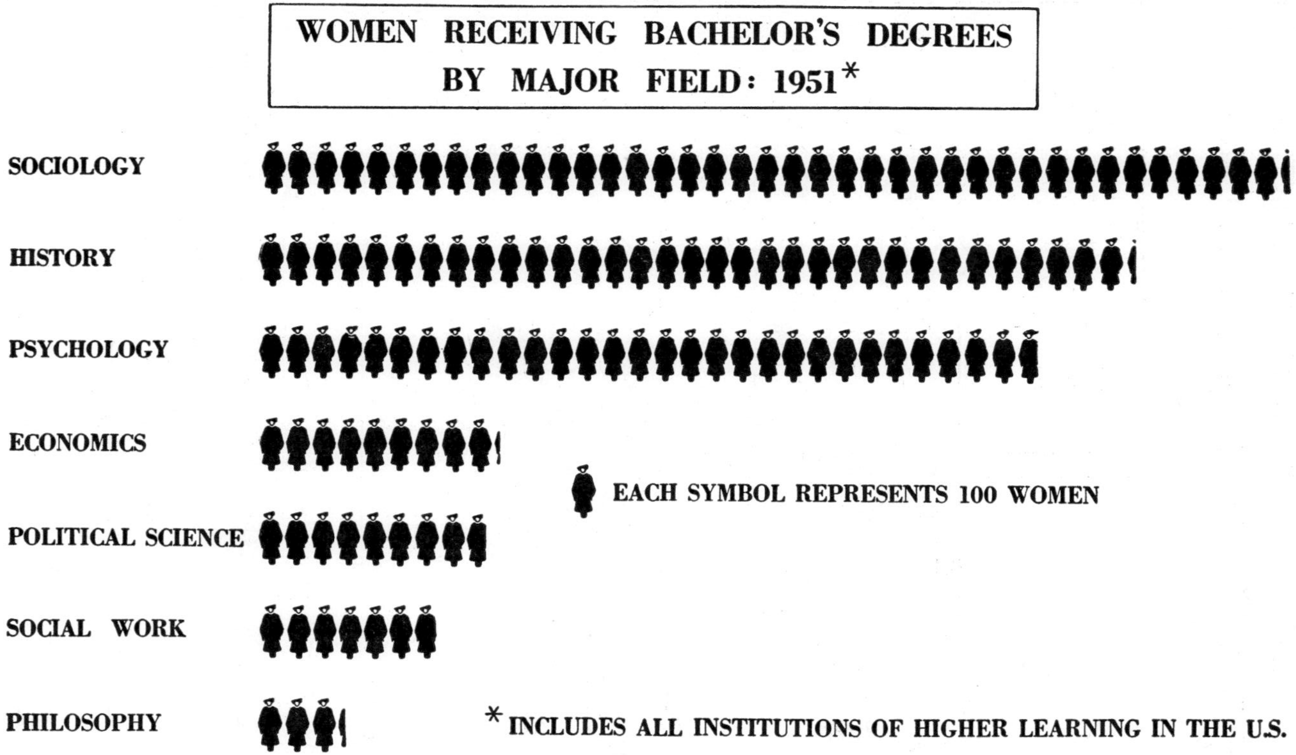

WOMEN RECEIVING BACHELOR'S DEGREES BY MAJOR FIELD: 1951*

SOCIOLOGY

HISTORY

PSYCHOLOGY

ECONOMICS

POLITICAL SCIENCE

SOCIAL WORK

PHILOSOPHY

EACH SYMBOL REPRESENTS 100 WOMEN

*INCLUDES ALL INSTITUTIONS OF HIGHER LEARNING IN THE U.S.

Fig. 3-22   Pictorial unit chart. It will be observed that this type of chart is very similar to the simple bar chart.

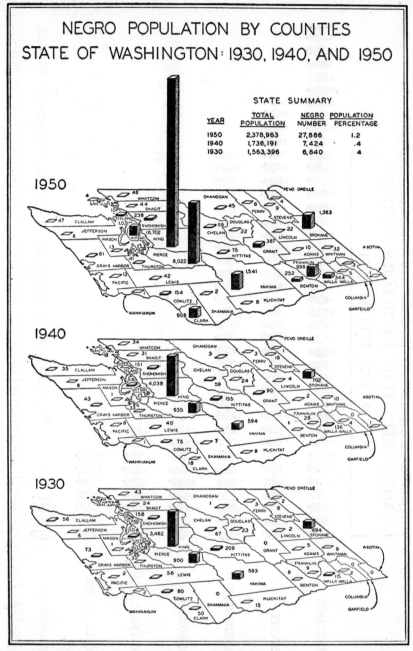

**Fig. 3-23 Maps with superimposed columns drawn in oblique projection.**
Charts drawn in three-dimensional form are useful in eliciting popular appeal,
since they possess depth and other pictorial qualities.

generally confined to popular presentation. *Pictorial charts* are seldom employed in technical presentations of research findings. Pictorial symbols may represent units of measurement, or they may function merely as artistic embellishment. Of the several types of pictorial charts the pictorial unit chart, which is illustrated in Fig. 3-22, is the most satisfactory and reliable. Each symbol is of uniform size and of specified value. There is a clear, logical relationship between the pictorial unit chart and the simple bar and column chart. Instead of divisions on a scale, pictorial symbols represent measurement values in the form of distinct units.

### 3-16   Three-dimensional Charts

In recent years it has become common practice to portray rectilinear coordinate graphs, pie charts, bar and column charts, maps, and other graphic forms in *three-dimensional projection*. Projection charts, with their depth and picturelike qualities, possess strong popular appeal. Three basic types of projection applicable to the construction of statistical charts are axonometric, oblique, and perspective. Figure 3-23 is an illustration of maps drawn in projection. The columns are in oblique projection and the base maps in quasi-perspective projection. It also will be observed that Fig. 3-21 has been drawn in projection form.

### 3-17   Summary

Rectilinear coordinates portray differences in amount.

Semilogarithmic coordinates show relative changes.

In a frequency polygon, frequencies are plotted at the midpoint of the class intervals, and the plotted points are connected by straight lines.

In a histogram, frequency is indicated by the area of a rectangle.

Cumulative frequency distributions of the "more than" or "less than" type are portrayed by ogives.

Linear presentations of statistical data are less ambiguous than areal or cubic representations.

In general, maps are the most effective technique for portraying spatial relations of statistical data. There are four basic types of statistical maps.

Because of their popular appeal, pictorial and three-dimensional charts are especially valuable for nontechnical presentation.

# 4 Basic Rules of Summation

**VOCABULARY**
constant $(a, b, c)$
variable $(X, Y, Z)$
summation $(\Sigma)$
number of cases $(N)$

This chapter is devoted to a discussion of certain rules of summation (adding). It is strictly preparatory. These rules should be clearly understood since they are basic to many of the techniques and procedures covered in various parts of the book. They are helpful in computing, as well as in deriving equations.

## 4-1 Variables and Constants

As two men were driving by a herd of cattle grazing in a pasture, one turned to the other and said, "Those 162 cows are fine stock."

His companion said, "How could you count those cows so quickly?"

"Easy. I counted the legs and divided by four."

Let us express this remarkable technique in algebraic terms. If $Y$ represents the number of cows and $X$ the number of legs, then

$$Y = \frac{X}{4}$$

As one herd after another is observed, the number of cows and the number of legs would both change, but the divisor 4 would remain the same.

We could use the number of eyes instead of the number of legs to be symbolized by $X$. Accordingly the equation would be

$$Y = \frac{X}{2}$$

The $X$ and $Y$ would change with respect to the number of cows in the field, but the divisor would always be 2. If we liked, our original equation could be written in such an abstract manner that it would include both the original technique as well as our adaptation. Thus the equation may be written as follows:

$$Y = \frac{X}{a}$$

where $Y$ = the number of cattle
     $X$ = the number of eyes or legs
     $a$ = the constant used, in this problem either 2 or 4

A *constant* does not change its numerical value within a set of data. A constant has a constant numerical value. In the two examples above, 2 and 4 are constants.

A *variable* can represent any one of a set of numerical values. Therefore, a variable varies in value. The number of cows, the number of legs, and the number of eyes all change from herd to herd, so they are all variables.

Let us use the first letters of the alphabet, $a$, $b$, and $c$, to symbolize (stand for) constants. This is arbitrary. If we wished, orange, lemon, and lime could be used to symbolize constants. All that is necessary is agreement concerning the meaning of the symbols. Just as arbitrarily, statisticians generally use the last letters of the alphabet, written in capitals, $X$, $Y$, and $Z$, to symbolize variables.

What does $aX$ mean? It means that some value of a variable, like age, is being multiplied by a constant, some number. Reference can be made to an $X$ without specifying what variable is being discussed, and the $a$ represents all possible constants by which the $X$ could be multiplied. This has certain advantages, since general statements can be made that are applicable to all variables and to all constants which are being multiplied together. This is an example of the major reason why algebraic notation is used. Substituting letters for particular numbers makes it possible to solve problems and apply the results to many specific cases which differ only in the values of the variable. Twice a person's age is half of four times his age, whether he is 3 months old or 999 years old. Twice a person's rating on friendliness is half of another person's rating if the other person had a score four times as high. Expressing these ideas in algebraic form

$$2X = \tfrac{1}{2}\,(4X)$$

## 4-2  Summation

Let us use the symbol $\Sigma$ (Greek capital sigma) to stand for the *summation* process, the adding together of all the cases of the term to the right of the $\Sigma$ sign. The expression $\Sigma X$ means to add up the numerical values of $X$, starting with the first case and ending with the last. If desired, each case can be designated separately. The first case will be referred to as $X_1$, the second $X_2$, the third $X_3$, and so forth. If it were agreed that $N$ stands for the total number of cases, the last case would be designated $X_N$.

Rewriting in symbols what has just been defined in words,

$$\Sigma X = X_1 + X_2 + X_3 + X_4 + \cdots + X_N$$

To illustrate, suppose we want to know the total number of years of schooling for a group of four people. They have gone to school for 14, 16, 11, and 15 years, respectively. Therefore, $X_1 = 14$, $X_2 = 16$, $X_3 = 11$, and $X_4 = 15$. Then

$$\Sigma X = 14 + 16 + 11 + 15$$
$$= 56$$

Sometimes the idea of summation is expressed in a manner which specifies the exact cases to be added. $X_i$ is the algebraic expression referring to a particular case. If $i$ were 2, then the reference is to the second case.

Therefore, $\displaystyle\sum_{i=1}^{N} X_i$ indicates that all the values of the variable from the first case to the last case, from $X_1$ to $X_N$, should be added. By the same reasoning if only the fifth to eighth cases are to be added, this operation would be written as

$$\sum_{i=5}^{8} X_i$$

Since no operations which require summation for only part of a group of cases are included in this elementary text, it is unnecessary to use such a complicated method of symbolizing summation. From this point on, since $\Sigma X$ has the same meaning as $\sum_{i=1}^{N} X_i$, $\Sigma X$ will be used to represent the summation of all the cases.

## 4-3    First Rule of Summation

*Rule* 1. The summation of the sum of two or more terms is equal to the summation of the separate sums of the terms. This rule sounds difficult, but in application it is very simple. Rule 1 can be expressed in symbolic form as follows:

$$\Sigma(X + Y) = \Sigma X + \Sigma Y$$

Again, another symbolic example of this rule is

$$\Sigma(Y - Z) = \Sigma Y - \Sigma Z$$

**Supplementary Explanation.** Some people have trouble in using parentheses in algebraic manipulation. Parentheses indicate that all material within the parentheses is to be treated as a single unit.

When two letters, or a number and a letter, are written next to each other, with no plus or minus sign intervening, then multiplication is called for. For example, $2X$ indicates that each value of variable $X$ be multiplied by 2; $XY$ specifies that each value of $X$ be multiplied by its corresponding value of $Y$; and $(X - 6)$ denotes a subtraction of 6 from each value of $X$.

The meaning of parenthetical expressions in such problems is similar. For example, $(X)(Y)$ is the same as $XY$, and $2(X)$ is identical to $2X$. In the expression $2(X + Y)$, the term $(X + Y)$ is considered a single unit. The multiplication by 2 applies to the sum of $X$ and $Y$. If $X$ were 3 and $Y$ were 7, then $2(X + Y)$ is equal to $2(3 + 7) = 2(10) = 20$.

On the other hand, $2(X) + Y$ requires that only the $X$ value be multiplied by 2, with the product added to the value of $Y$.

**Illustration.** Find $\Sigma(X + Y - Z)$. If the long method were applied to this problem, it would be necessary to add together the $X$ and $Y$ values for each case, and then subtract the corresponding $Z$ value for each case, after which the results of the additions and subtractions are summated. The step-by-step computational procedure is illustrated in the following table:

| Case number | $X$ | $Y$ | $Z$ | $X + Y$ | $X + Y - Z$ |
|---|---|---|---|---|---|
| 1 | 9 | 8 | 11 | 17 | 6 |
| 2 | 2 | 4 | 7 | 6 | −1 |
| 3 | −5 | 6 | 3 | 1 | −2 |
| 4 | 8 | 15 | 6 | 23 | 17 |
| | $\Sigma X = 14$ | $\Sigma Y = 33$ | $\Sigma Z = 27$ | | $\Sigma(X + Y - Z) = 20$ |

In problems of this kind, Rule 1 provides a relatively quick method for obtaining identical results. The saving in time is not very apparent in this example, since there are only a few cases. When there are many cases, however, this method is the fastest and easiest.

Let us apply Rule 1 to the same problem:

$$\Sigma(X + Y - Z) = \Sigma X + \Sigma Y - \Sigma Z$$
$$\Sigma X = +19 - 5 = 14$$
$$\Sigma Y = 33$$
$$\Sigma Z = 27$$
$$\Sigma X + \Sigma Y - \Sigma Z = 14 + 33 - 27 = 20 \quad \text{(the same answer as above)}$$

## 4-4   Second Rule of Summation

*Rule 2.* The summation of a constant times a variable is equal to the constant times the summation of the variable. In simpler terms, if we have a variable that is always multiplied by the same number, the summation can be found quickly by adding the values of the variable and multiplying the sum by the constant. Rule 2 may be symbolized in the following manner:

$$\Sigma aX = a\Sigma X$$

**Illustration.** Given the following data, $\Sigma aX$ can be derived according to the long method by multiplying each value of $X$ by the constant $a$ and summing the results:

| Case number | $X$ | $a$ | $aX$ |
|---|---|---|---|
| 1 | 11 | 4 | 44 |
| 2 | 23 | 4 | 92 |
| 3 | −8 | 4 | −32 |
| 4 | 70 | 4 | 280 |
| 5 | 5 | 4 | 20 |
| | $\Sigma X = 101$ | | $\Sigma aX = 404$ |

By using the short method, however, the same results can be obtained by a much simpler and more efficient procedure.

$$\Sigma X = +109 - 8$$
$$= 101$$
$$a\Sigma X = (4)(101)$$
$$= 404$$

## 4-5   Third Rule of Summation

*Rule* 3. The summation of a constant is equal to the constant times the number of cases. Indicated symbolically,

$$\Sigma a = Na$$

**Illustration.** Find the total number of hands in a group of eight people. The number of hands possessed by different people is usually a constant, 2. We could find the number of hands in the group by adding the constant eight times: $2 + 2 + 2 + 2 + 2 + 2 + 2 + 2 = 16$.

Rule 3 states that the total number of hands could be found by multiplying the constant (2) times the number of cases (8).

$$\Sigma a = Na = 8(2) = 16$$

## 4-6   Misapplication of Summation

The basic rules for summation now have been explained. Before testing your ability in applying them, some words of caution may be necessary. Sometimes students add rules of their own which are not correct. For example, they may consider that $\Sigma X^2 = (\Sigma X)^2$ and that $\Sigma XY = (\Sigma X)(\Sigma Y)$. Both these statements are false.

$\Sigma X^2$ means that each value of $X$ is squared (multiplied by itself) and all these results are added together:

$$\Sigma X^2 = X_1^2 + X_2^2 + X_3^2 + \cdots + X_N^2$$

For example, suppose the numbers 2, 4, and 7 are substituted in the formula. Then

$$\Sigma X^2 = \Sigma(2^2 + 4^2 + 7^2)$$
$$= \Sigma(4 + 16 + 49)$$
$$= 69$$

Now let us examine the meaning of $(\Sigma X)^2$. This specifies that all the values of $X$ are added, and then the sum is squared (multiplied by itself). For example, if $X$ assumes the same values 2, 4, and 7 as in the foregoing series, then $(\Sigma X)^2$ becomes $(2 + 4 + 7)^2 = (13)^2 = 169$. Notice that, by using the same values of $X$, $(\Sigma X)^2$ gives very different results than $\Sigma X^2$.

A similar source of confusion is the interpretation of $\Sigma XY$. Students sometimes assume that $\Sigma XY$ can be obtained by multiplying $(\Sigma X)$ by $(\Sigma Y)$. $\Sigma XY$ indicates that for every case the $X$ value of that case is multiplied by the corresponding $Y$ value of that case and the products thus derived are summed.

**Illustration.** Find the total wages of men who work at different hourly rates of pay. If $X$ is the pay rate and $Y$ the hours of work, then $XY$ is the amount each earns. $\Sigma XY$ is the total wages of the group.

| Case number | X (pay rate in dollars) | Y (hours worked) | XY (pay received in dollars) |
|---|---|---|---|
| 1 | 5 | 2 | 10 |
| 2 | 3 | 3 | 9 |
| 3 | 2 | 7 | 14 |
| 4 | 1 | 15 | 15 |
| | | | $\Sigma XY = 48$ |

It readily can be observed that $(\Sigma X)(\Sigma Y)$ would give a different and, of course, incorrect answer. $\Sigma X = 11$, and $\Sigma Y = 27$. By multiplying the two sums together $(11)(27)$, the product would be 297, a figure markedly different from that indicated by $\Sigma XY$.

**Problems. 1.** Rewrite the following expression in a form that would be more adaptable to rapid calculation:

$$\Sigma(X^2 + 15X - 5)$$

*Ans.* $\Sigma X^2 + 15\Sigma X - 5N$

**2.** Rewrite the following in a form which would aid in computation:

$$\Sigma(6X^2 + 5XY + 11)$$

$$Ans.\ 6\Sigma X^2 + 5\Sigma XY + 11N$$

## 4-7   Summary

1. $\Sigma(X + Y - Z) = \Sigma X + \Sigma Y - \Sigma Z.$
2. $\Sigma aX = a\Sigma X.$
3. $\Sigma a = Na.$
4. $(\Sigma X)^2$ does not equal $\Sigma X^2.$
5. $\Sigma XY$ does not equal $(\Sigma X)(\Sigma Y).$

## PROBLEM FOR CHAP. 4

**1.** A scale is developed for measuring the degree to which individuals feel that they are an integral part of a particular group. The scores are expressed in decimals, both positive and negative. These scores are added together in order to derive a measure of cohesiveness for the entire group.

To improve the scale as a working tool, it is desired to eliminate the decimals and negative numbers. Therefore, each individual score is multiplied by 100, and 200 then added to the product. Expressed symbolically,

Group cohesiveness $= (100X_1 + 200) + (100X_2 + 200) + \cdots$
$$+ (100X_n + 200)$$

which can be summarized as $\Sigma(100X_i + 200)$.

If the sum of the original scores for a group was 3.20 when there were 10 persons in the group, what will be the sum of the transformed scores?                    *Ans.* 2,320

# 5 Measures of Central Value

If we have a large amount of numerical data, the first step is to summarize the data by indicating some of its more salient characteristics. One of the most important summarizing measures in statistical description is the average or central value. This chapter is concerned with definitions and applications of different kinds of measures of central value.

## 5-1 Main Divisions of Statistics

Statistics, which deals with the analysis of numerical data, can be separated into two main divisions: *descriptive statistics* and *statistical inference* (often called "inductive statistics").

Descriptive statistics attempts to condense and summarize quantitative data in a clear and convenient form. For example, grades of a class in English may be summarized by calculating the average grade or by indicating the range of grades from lowest to highest. This summary would be a description of the data without any attempt to derive broader generalizations.

Statistical inference is concerned with deriving broad, inclusive generalizations. For example, if an attempt is made to estimate the English grades for all the children in a school from the grades of one class, this would be considered a problem in statistical inference.

The first part of this book is concerned only with descriptive statistics, and the second part with statistical inference or inductive statistics. As indicated in the introductory paragraph, this chapter is devoted to only one kind of summarizing measure, or different techniques for describing an average or central value.

## 5-2    Ambiguity of the Term "Average"

As used in everyday speech, "average" is not a clearly defined concept. When we say, "The average man makes $2,000," and "The average income of a group of men is $3,000," these statements may not necessarily be contradictory. Both these statements may be correct if the meaning of "average" has shifted. Ordinarily, in statistics we do not use the vague term "average" unless the meaning is clearly stated. In order to avoid ambiguity, from this point on we shall carefully define each type of central value or average.

## 5-3    Arithmetic Mean

The *arithmetic mean* is the most widely used measure of central value. Although there are several kinds of means, the arithmetic mean is used so often that a reference to the "mean" will always refer to the arithmetic mean. The mean of a variable is indicated by placing a line over the symbol representing the variable. For example, $\bar{X}$ is the mean of the $X$ variable, and $\bar{Y}$ is the mean of the $Y$ variable. The mean is calculated by dividing the sum of all the values of a variable by the number of cases. Stated in algebraic form,

$$\bar{X} = \frac{\Sigma X}{N}$$

## 5-4    Sum of Deviations from Mean

An important property of the mean is that the sum of the deviations from the mean is always equal to zero. One way of indicating the deviation of the value of some case from the mean is by the algebraic expression, $(X - \bar{X})$. If, for example, the item has a value of 20 and the mean is 16, then according to the expression $(X - \bar{X})$ the deviation would be 4. If $X$ were 9 and $\bar{X}$ were 11, the deviation would be $-2$, indicating that the item is 2 units below the mean. The sum of negative deviations from the mean will always be exactly equal to the sum of the positive deviations, thus giving a total of zero.

**Proof.** It readily can be proved that the sum of the deviations around the mean is necessarily zero. The proof merely requires a simple application of two rules of summation. We want to demonstrate that $\Sigma(X - \bar{X}) = 0$.

It will be recalled from Rule 1 in the foregoing chapter that

$$\Sigma(X - \bar{X}) \text{ can be written } \Sigma X - \Sigma \bar{X}$$

The mean of a particular set of numbers is just one number, a constant for that set of numbers. For that group of numbers, the mean is fixed and cannot vary. According to Rule 3,

$$\Sigma X - \Sigma \bar{X} = \Sigma X - N\bar{X}$$

Furthermore, by definition, $\bar{X} = \Sigma X/N$

Accordingly, $\Sigma X/N$ can be substituted for $\bar{X}$ at any time without changing the numerical result. Let us do this in $\Sigma X - N\bar{X}$, which equals the sum of the deviations around the mean. Thus,

$$\Sigma X - N\bar{X} = \Sigma X - N\left(\frac{\Sigma X}{N}\right)$$

The sum of the deviations around the mean is now equal to $\Sigma X - N(\Sigma X/N)$. Carefully examine the last term. We are going to add the values of all the items, divide the total by the number of items, and also multiply by the number of items. By both multiplying and dividing by the number of cases, the original $\Sigma X$ will result.

$$N\left(\frac{\Sigma X}{N}\right) = \Sigma X$$

If $\Sigma X$ is substituted for the second term, the sum of the deviations around the mean becomes $\Sigma X - \Sigma X$.

Subtracting any number from itself leaves zero, so that the sum of the deviations around the mean is always equal to zero.

It is awkward always to write $(X - \bar{X})$ for a deviation from the mean. Let us use $x$, a small italicized letter, to stand for $(X - \bar{X})$, a deviation from the mean. This introduces no essential change. For example, it already has been shown that $\Sigma x = 0$. Needless to say, it is of great importance to note the difference between $X$, the value of some case of a variable, and $x$, the deviation of that case from the mean.

## 5-5  Sum of Squared Deviations from Mean

The deviations around the mean have another interesting property. The sum of the squared deviations around the mean $(\Sigma x^2)$ is less than the sum of the squared deviations around any other value. Another way of expressing the same idea is as follows: the sum of the squared deviations is a minimum when taken around the mean.

**Illustration.** Find $\bar{Y}$, $\Sigma y$, and $\Sigma y^2$.

| $Y$ | $y$ | $y^2$ |
|---|---|---|
| 2 | $-3$ | 9 |
| 9 | 4 | 16 |
| 7 | 2 | 4 |
| 2 | $-3$ | 9 |
| $\Sigma Y = 20$ | $\Sigma y = 0$ | $\Sigma y^2 = 38$ |

To find $\bar{Y}$, all we need is the sum of the $Y$ values and the number of cases.

$$\bar{Y} = \frac{\Sigma Y}{N} = \frac{20}{4} = 5$$

The deviations from the mean are determined by subtracting the mean from the value of each case. The first case has a value of 2, and subtracting the mean, 5, gives a deviation from the mean of $-3$. This is done for each item. The sum of the positive deviations is 6, and the sum of the negative deviations is also 6. Adding them gives a sum of zero. Expressed symbolically,

$$\Sigma y = 0$$

To obtain $\Sigma y^2$, we must square each deviation from the mean (multiply each deviation by itself) and sum the squared deviations. It will be observed from the above illustration that $\Sigma y^2 = 38$.

**Supplementary Explanation.** Deviations from the mean are both positive and negative, yet squared deviations are all positive. This is an illustration of the common algebraic rule for determining the sign ($+$ or $-$) of a product or quotient. When multiplying or dividing, two like signs (either positive or negative) always give a plus sign, while two unlike signs give a minus sign. When squaring a number, the sign of the two numbers is the same, so the sign must be positive.

**Illustration**

$$(-2)(4) = -8$$
$$(-2)(-2) = 4$$
$$(3)(7) = 21$$
$$(8)(-5) = -40$$

When *multiplying or dividing* three or more numbers, the rules still hold. One merely considers some pair of numbers, then the result is multiplied or divided by another number, that result by the next number, and so on. For example, $(-7)(-2)(-1)(2) = -28$. In deriving this answer, the first step is to multiply $-7$ by $-2$, which gives 14; next $(14)(-1)$ is $-14$; and $(-14)(2)$ equals $-28$.

Another way of obtaining the same result is to count the number of minus signs. If there is an odd number of minus signs (1, 3, 5), then the result will have a minus sign. In the example above, there are three minus signs. Three is an odd number, so the answer is negative.

When *adding or subtracting a fraction* (a fraction is a division problem), the same general rule applies. For example,

$$6 - \frac{-4}{2} = 8.$$

Solving this problem step by step, it will be seen that $-4$ divided by 2 equals $-2$, which is in turn multiplied by the minus sign in front of the fraction. This result is 2, which, added to 6, gives the answer, 8.

Continuing our illustration, let us show that $\Sigma y^2$ is a minimum around the mean. The sum of the squared deviations around the mean is 38, and the sum of the squared deviations will be larger if taken around any other value. Let us choose some number, say 6, and find the sum of the squared deviations about that number. The deviations are −4, 3, 1, and −4, respectively. Accordingly, the sum of the squared deviations is $16 + 9 + 1 + 16 = 42$, a number larger than 38. Additional experimentation would demonstrate that any other value would have a sum of squared deviations larger than the sum of squared deviations around the mean.

**Problem.** Find the $\bar{X}$, $\Sigma x$, and $\Sigma x^2$ of the following series:

$X$ (the number of children under 14 years of age in each family)

9

3

5

0

13                                        *Ans.* 6; 0; 104

### 5-6   Mean and Extreme Values

The mean should not be used unless full emphasis is desired for the extreme cases in a group. For example, we might be studying the mean income of a small group. Five men have incomes of $1,000, $2,000, $5,000, $4,000, and $60,000. Their mean income is $72,000/5, or $14,400. The presence of one wealthy man has produced a mean income which is not a very "typical" value. In fact, not a single member of the group has an income which is even close to the mean.

### 5-7   Median

The *median* is the second important measure of central tendency. It is that value which has as many cases below it as there are cases above it. There are just as many cases with values below the median as there are cases with values above the median. (See Sec. 8-4 for a more precise definition.)

The median of the distribution with values 2, 3, 7, 19, and 30 is 7. The number 7 has two cases below it and two above it. As you can see, finding the median is easy once the data are classified in the form of an array, that is, in order from lowest to highest or highest to lowest.

**Illustration.** What is the median income of men who have the following incomes: $8,000, $4,000, $3,000, $4,000, $100,000, $5,000, $7,000?

Arranging the incomes in an array,

$100,000
8,000
7,000
5,000
4,000
4,000
3,000

The median is $5,000, which has an equal number of cases on either side of it.

Both the preceding illustrations have an odd number of cases. When there is an even number of cases, a slight change must be made in computing the median. For example, let us find the median of a distribution with an even number of cases, such as 1, 2, 4, 5, 7, 8. There are six cases. The number 4 has two cases below it and three cases above it, so by definition it cannot be the median. The number 5 has three cases below it and two cases above it, and therefore the median is not 5. But any number between 4 and 5 will satisfy the definition of the median. It will be seen that 4.1 has three cases above and three below, and the same is true of 4.8.

In order to obtain consistent results in problems of this type, a well-established arbitrary rule must be followed. When there is an even number of cases, the median is the value halfway between the two central numbers. For this example, the central numbers are 4 and 5, so the median is 4.5.

**Problem.** Find the median for the following distributions:

4, 2, 3, 1, 7
1, 2, 5, 4
1,000; 2,000; 4,000; 2,000
7; 8; 9; 1,000,000

*Ans.* 3; 3; 2,000; 8.5

## 5-8  Median and Extreme Values

Unlike the mean, the median is not greatly affected by extreme values. In the computation of the median, the crucial question is the number of cases above and below the median. One unit above the median has exactly the same influence as a million units above the median.

The relative impact of a single item on the mean and median can be observed readily by adding an extreme case to a series. For example, the ratings in self-confidence of four persons are 64, 70, 72, and 68. The mean for the group is 68.5, and the median, 69. If the score for the leader of the group with a rating of 93 is added to the series, then the mean self-confidence score will rise to 73.4, while the median will increase to only 70.

## 5-9  Mode

The *mode* is another measure of central value, but it is used less frequently than either the mean or the median. The mode is the most frequently occurring value. In a distribution with the values 11, 19, 37, 21, 22, 11, 14, 21, 23, 21, 18, 30, and 24, the mode is 21. It is 21 because there are more cases with that value than any other value.

The mode is not affected at all by extreme cases. For example, if a case with the value 100 were added to the distribution in the preceding paragraph, the mode still would be 21.

Some distributions have two or more frequently occurring values that are widely separated. The mode is not a good measure of central value in such cases.

**Problem.** Find the modal value for the following sets of data:

> 1, 3, 5, 5, 7, 11, 11, 11, 13, 14, 14, 18, 22, 24
> 8, 11, 14, 19, 22, 23, 26, 30
> 4, 6, 19, 6, 8, 23, 8, 14, 15
> 1, 5, 5, 7, 9

*Ans.* 11; no mode; 6 and 8; 5

## 5-10  Geometric Mean

There are other measures of central tendency which are occasionally used. Only one of these will be discussed here.

It is considered mainly because it will be used in a later chapter.

The *geometric mean* is the $N$th root of the product of $N$ numbers. The meaning of this seemingly formidable definition can best be made clear by several illustrations.

Suppose we wish to find the geometric mean of the following numbers: 2, 4, and 8. The product of these numbers is 64, since 2 times 4 is 8, and 8 times 8 is 64. The geometric mean indicates that we must find the third root (cube root) of these numbers, since $N$ (number of cases) is 3. In other words, we seek some number, $a$, such that, when three $a$s are multiplied together, we obtain 64. In this illustration $a$ is equal to 4 because $(4)(4)(4)$ is 64. Thus 4 is the geometric mean of 2, 4, and 8.

**Illustration.** Find the geometric mean of 4 and 9.

The product of these numbers is 36. We want a number, $a$, such that $a$ times $a$ (two $a$s because there are two original numbers) will be 36. It can be seen readily that 6 times 6 is 36 (6 is the square root of 36), which indicates that 6 is the geometric mean of 4 and 9.

**Problem.** Find the geometric mean of 3 and 27.                    *Ans.* 9

## 5-11  Effect of Skewness on Mean and Median

In Fig. 5-1 the two most widely used measures of central tendency are plotted on several frequency polygons. The relative influence of skewness is obviously greater for the mean than for the median. Each measure provides a way of summarizing the values of a frequency distribution on the $X$ axis. The mean and median are typical measures which communicate specific facts about frequency distributions. Some information is unavoidably lost when a single measure is used to represent an entire distribution, but there is a gain in simplicity of expression and ability to compare distributions with one another.

This chapter has treated somewhat artificial problems using ungrouped data. In Chap. 8 the methods of computation using grouped data will be explained. At this point, it is only necessary to have a general view of the summarizing measures of central value.

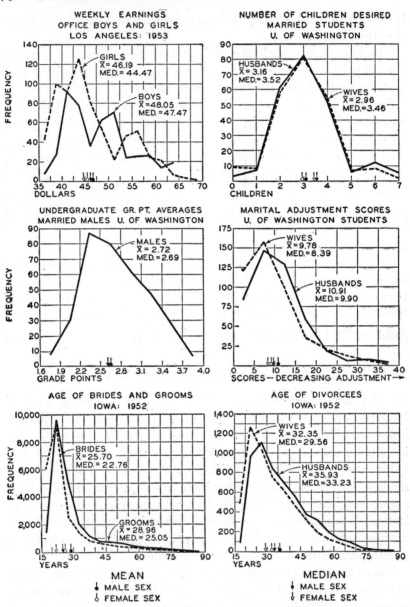

**Fig. 5-1** Graphic illustrations of mean and median. It will be observed that both the mean and the median represent values on the $X$ scale.

## 5-12   Summary

$$\bar{X} = \frac{\Sigma X}{N}.$$

The median has as many cases below it in value as above it in value.

The mode is the most frequently occurring value.

The sum of the deviations around the mean is zero.

The sum of the squared deviations around the mean is a minimum.

The mean is greatly affected by the introduction of one or more extreme cases, the median affected very little, and the mode not affected at all.

## PROBLEMS FOR CHAP. 5

1. A mother is asked to predict the responses of each of her five children to a series of 10 intimate questions. The number of errors she makes is, respectively, 6, 4, 3, 5, and 9.

   *a.* What is the mode?                           *Ans.* No mode

   *b.* What is the median?                          *Ans.* 5

   *c.* What is the mean?                            *Ans.* 5.4

   *d.* What is the deviation of each case from the mean?

                       *Ans.* .6, $-1.4$, $-2.4$, $-.4$, and 3.6

   *e.* What is the sum of the deviations around the mean?     *Ans.* 0

   *f.* What is the sum of the squared deviations around the mean?

                                    *Ans.* 21.20

2. A small community of 1,000 persons becomes the site of a new factory, with its population rising in 10 years to 25,000 inhabitants. What is the geometric mean of the two populations?      *Ans.* 5,000

3. A teacher is interested in the effect of television on schoolwork. She separates the class into two groups on the basis of their grades. The 11 better students watch television for 6, 14, 19, 29, 12, 3, 0, 7, 9, 8, and 10 hours a week. The 9 poorer students spend 8, 11, 3, 15, 25, 7, 16, 5, and 9 hours in front of the television screen.

   *a.* Which group has the higher mean?     *Ans.* The poorer students

   *b.* Which group has the higher median?

                        *Ans.* Both medians are identical

# 6  Measures of Variability

In the preceding chapter measures of central value were discussed. No single measure, however, can reveal everything about a set of data. Two sets of data having identical means may be very different in other respects (Fig. 6-1). For example, if Adolf

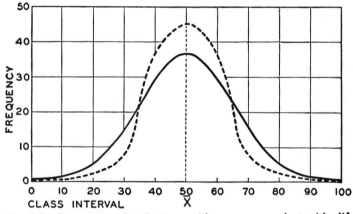

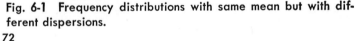

Fig. 6-1  Frequency distributions with same mean but with different dispersions.

obtains a score of 100 in one test and 50 in another, his mean grade will be the same as Benito's, who receives 75 in each of the two tests. However, Adolf is more variable in his scores than is Benito. This chapter will present a discussion of measures of variability or dispersion. The combination of a measure of central value and a measure of variability will help to provide a more complete and more meaningful characterization of the data.

## 6-1    Range

A very young child is playing by the side of a small brook. "Isn't that dangerous?" you ask your host. "Oh, no," he replies, "the mean depth is only 8 inches." That stream may be safe, but your host's answer does not settle the issue. It is possible that the water is only a few inches deep in most places but that there are a few spots where the depth may be over the child's head.

The *range*, defined as the difference between the largest and smallest value, is a measure of variability which utilizes only extreme values. In the case of the brook above, if the most shallow spot were 1 inch and the greatest depth 11 inches, the range would be 10 inches. If your host had said that the mean was 8 inches and the range 10 inches, the stream's depth would no longer worry you. Of course, a statement of the maximum depth also would have satisfied you.

**Illustration.** A group of children are invited to a birthday party. The mother wants to plan some games. She notes that the ages of the children are 7, 2, 9, 3, 8, and 7. The mean age is 6, but the range is 7 (9 minus 2). The mother decides that the range is too great for all the children to enjoy the same games.

**Problems. 1.** A group of students in social statistics take a test of mathematical knowledge. Their scores are

11, 27, 44, 36, 72, 81, 19, 39, 56, 44, 48, 60, and 17

Find the range. *Ans.* 70, or 11 to 81

**2.** These same students take a test concerned with knowledge of the positions of important political figures. Their scores:

88, 79, 83, 91, 43, 87, 21, 90, 82, 79, 65, 77, and 84

Find the range.                                        *Ans.* 70, or 21 to 91

## 6-2   Advantages and Disadvantages of Range

The advantage of the range as a measure of variability is its ease
of computation. Subtracting the lowest value from the highest
value is so simple that quick and accurate results will be obtained
by inexperienced persons. On the other hand, the range has
several major deficiencies which limit its usefulness.

Disadvantages of the range as a measure of variability are:

*a.* The range is completely dependent on the two extreme
cases. The highest and lowest values in the distribution determine the range, and all other values have absolutely no influence.
In each of the two preceding problems, the range was 70,
although the test scores on mathematical knowledge were really
much more variable than the test scores on political personalities. Ten of the scores for the test on political personalities are
between 77 and 91, while no such bunching is found in grades
on the mathematics test. This illustrates the fact that the range
may not be an adequate measure of variability.

*b.* The dependence of the range on the two extreme values
makes it a very unstable measure. The presence or absence of a
single extreme score can exercise a marked influence on the
range. For example, if the person who received a score of 21
in knowledge of political figures had been ill and missed the test,
the range would have been 48 (91 minus 43) instead of 70.

*c.* The range also is affected by the number of cases in the
distribution. Generally, the larger the number of cases, the more
likely one is to find extreme values. If the mathematical-
knowledge test were administered to 10,000 students in social
statistics, rather than merely 13, the likelihood of finding
students with scores near 100 and 0 would increase considerably.
The range in the sample of 10,000 persons would approach 100.
The greater range for 10,000 students might give the erroneous
impression that there has been a general increase in the variability of scores.

## 6-3    10-90 Percentile Range

Let us consider another measure of variability which, like the range, is based on the difference between high and low values. It is unlike the range, however, in that it is not dependent on the two most extreme cases. To understand this second measure of variability, it is necessary to explain the meaning of *percentile*. For example, the 12th percentile is that value below which there are 12 per cent of the cases. Again, if 40 per cent of the students in a class have a score below 78.5, then 78.5 is the 40th percentile. By recalling the discussion of the median, its relationship to the concept of percentile can be observed readily. The median is always the 50th percentile, since the median has half the cases above it and half below it. When percentiles are used, the proportion of cases below certain points can be determined. Since the median has half the cases below it, it is the 50th percentile (50 per cent is equal to $\frac{1}{2}$). The $j$ percentile has $j$ per cent of the values below it, to express the idea abstractly.

**Supplementary Explanation.** The word *cent* in French means "hundred" (Latin, *centum*), and this same idea is present in the phrase "per cent" (Latin, *per centum*), meaning "per hundred." For example, 14 per cent means 14 out of every 100. If you say that 10 per cent of the soldiers are showing symptoms of battle fatigue, you mean that 10 out of every 100 show such symptoms.

The notion of per cent can be written in several ways. The following symbols all mean the same thing: 24%, 24 out of 100, .24, $\frac{24}{100}$, twenty-four hundredths, twenty-four per cent.

One hundred per cent, of course, includes the entire group. If 100 per cent of a class of 120 volunteer for an experiment, then 120 persons have volunteered. If 40 per cent of the class volunteer, then forty-hundredths of the entire group have volunteered. Let us compute how many volunteers there were. Forty per cent of 120 volunteered. That is the same as 120 multiplied by .40.

Thus,

$$
\begin{array}{r}
120 \\
.40 \\
\hline
000 \\
480 \\
\hline
48.00
\end{array}
$$

It will be observed that there are two figures on the right of the decimal point in the two numbers being multiplied together, so it must be made certain that two places are marked off to the right of the decimal in the answer. Accordingly, the decimal is placed between the 8 and the zero, and the answer is 48 volunteers.

**Problem.** How many scores are below the 30th percentile in a group of 90 scores on interest in world government?                    *Ans.* 27

In some respects, the 10-90 *percentile range* is similar to the range as a measure of variability. It is found by subtracting the 10th percentile (the value with 10 per cent of the cases below it) from the 90th percentile (the value with 90 per cent of the cases below it).

**Illustration.** An investigator is attempting to study the attitude of medieval societies toward capital punishment. He counts the number of crimes punishable by death according to law in each society. The results for 30 medieval societies are as follows:

8, 11, 2, 13, 1, 6, 22, 0, 15, 46, 21, 18, 8, 15, 32, 7, 17, 5, 14, 6, 7, 0, 11, 8, 9, 4, 10, 9, 6, 5

In an ungrouped series of this kind, the first step in locating percentiles is to arrange the data in the form of an array:

0, 0, 1, 2, 4, 5, 5, 6, 6, 6, 7, 7, 8, 8, 8, 9, 9, 10, 11, 11, 13, 14, 15, 15, 17, 18, 21, 22, 32, 46

To determine the 10-90 percentile range, we must find the 10th and 90th percentiles. The 10th percentile should have 10 per cent of the values below it. There should be 3 cases below the 10th percentile, since there are 30 cases and 10 per cent of 30 is 3. The value which is the 10th percentile is 1.5, since it has 3 cases below it. It will be observed from the array that any number between 1 and 2 has 3 cases below it, and the choice of the number halfway between (1.5) is an arbitrary rule whose basis will be more fully understood after rereading Sec. 2-6.

The 90th percentile should have 27 cases below it, 90 per cent of 30. The value 21.5 is the 90th percentile. The 10-90 percentile range, therefore, is 20, the difference between 1.5 and 21.5.

**Problem.** Find the 10-90 percentile range for the following scores on a scale designed to measure sociability:

2, 3, 4, 7, 9, 11, 18, 19, 21, 23, 25, 28, 28, 29, 31, 32, 32, 34, 35, 39

*Ans.* 31

## 6-4    Advantages and Disadvantages of 10-90 Percentile Range

In comparison with the total range, the 10-90 percentile range has the following advantages:

*a.* It is not based solely on the two most extreme measures, but depends on the values of the extreme 20 per cent of the cases, 10 per cent on each side, and the middle 80 per cent of the cases.

*b.* It is more stable than the range, less responsive to the inclusion or exclusion of a few cases. Chance fluctuations are less influential in its determination. Other measures of variability to be studied in this chapter will be even more stable than the 10-90 percentile range.

*c.* The 10-90 percentile range is not affected by the number of cases in the group under observation. Therefore, it shows no tendency to grow larger when more cases are included in the sample.

The disadvantages of the 10-90 percentile range are as follows:

*a.* Although fairly simple to compute, it is more difficult than the range.

*b.* It is comparatively unstable, but more stable than the range, since it is less affected by the two extreme cases.

## 6-5    Quartiles

Just as the 10-90 percentile range was developed as a more stable measure than the range, we shall proceed to discuss another measure of variability. This measure is based on *quartiles.* The concept of quartiles is very similar to the concept of percentiles. The first quartile, called $Q_1$, is that value below which there are one-quarter ($\frac{1}{4}$) of the cases. The second quartile, $Q_2$, has below it two-quarters or one-half ($\frac{2}{4} = \frac{1}{2}$) of the cases. The third quartile, $Q_3$, has three-quarters ($\frac{3}{4}$) of the cases below it. These three quartiles divide the distribution into four parts with an equal number of cases in each part.

**Illustration.** Find $Q_1$, $Q_2$, and $Q_3$ for the following distribution of the number of felonies that 20 apparently law-abiding males stated they had committed without being detected:

0, 0, 4, 4, 5, 6, 7, 8, 8, 8, 9, 9, 10, 10, 10, 11, 14, 15, 21, 23

$Q_1$ must have five cases below it ($\frac{1}{4}$ of 20), so $Q_1 = 5.5$.
$Q_2$ must have ten cases below it ($\frac{1}{2}$ of 20), so $Q_2 = 8.5$.
$Q_3$ must have fifteen cases below it ($\frac{3}{4}$ of 20), so $Q_3 = 10.5$.

The point $Q_2$, which has half the cases below it in value, is the median. Therefore, the median also can be labeled the second quartile or the fiftieth percentile.

Sometimes the term "quartile" is used to designate a group of cases rather than to the point that separates the groups of cases. More concretely, the first quartile sometimes means all the cases up to $Q_1$, the second quartile all the cases between $Q_1$ and $Q_2$, the third quartile comprises all the cases between $Q_2$ and $Q_3$, and the fourth quartile includes all the cases above $Q_3$. When one notes that the person scoring 15 in the illustration above is in the top quartile, the term is being used in this way. This confusion in terminology fortunately poses few difficulties. It is almost always clear which meaning of the term is intended.

## 6-6   Semi-interquartile Range, or Quartile Deviation

We are now ready to understand the next measure of variability. It has two common names, the *semi-interquartile range* and *quartile deviation*. The symbol for this measure is $Q$. It is found by subtracting $Q_1$ from $Q_3$ and dividing the result by 2. Expressed in a formula,

$$Q = \frac{Q_3 - Q_1}{2}$$

Using the data from the illustration above, $Q_1$ is 5.5, and $Q_3$ is 10.5. Therefore,

$$Q = \frac{10.5 - 5.5}{2}$$

$$= \frac{5}{2}$$

$$= 2.5$$

The quartile deviation is always equal to the mean distance that $Q_1$ and $Q_3$ diverge from the median, $Q_2$. Using the illustration again as a source of data, $Q_1$, with a value of 5.5, is 3 units away from the median, which is equal to 8.5. $Q_3$, at 10.5, is 2 units away from the median. The mean distance away is 2.5, the mean of 3 and 2. This is the same answer we obtained by using the other definition of the quartile deviation.

**Proof.** It can be demonstrated that the two ways of looking at the quartile deviation or semi-interquartile range always will give the same result. Let us express in algebraic form the idea that $Q$ is equal to the mean distance that $Q_1$ and $Q_3$ diverge from $Q_2$. The distance between $Q_1$ and $Q_2$ is $Q_2$ minus $Q_1$, since $Q_2$ is larger in value. The distance between $Q_3$ and $Q_2$ is $Q_3$ minus $Q_2$, since $Q_3$ is larger.

Now the mean of the two distances is found by adding them and dividing by 2. Expressing this idea in a formula,

$$Q = \frac{(Q_3 - Q_2) + (Q_2 - Q_1)}{2}$$

According to this formula, $Q$ is the mean of the distance that $Q_1$ and $Q_3$ differ from the median. By removing the parentheses,

$$Q = \frac{Q_3 - Q_2 + Q_2 - Q_1}{2}$$

Since $Q_2$ is being added and subtracted, the two operations cancel each other, and $Q_2$ can be dropped from the formula. This results in

$$Q = \frac{Q_3 - Q_1}{2}$$

This is the same formula given in the original definition of quartile deviation, showing that the two definitions are equivalent.

## 6-7  Mean Deviation

The *mean deviation*, the next measure of dispersion, is more stable than the quartile deviation. As its name implies, it is the mean of the deviations (irrespective of signs) from some measure of central value, usually the mean. In this text, the mean deviation will always be around the mean. In Chap. 5 it was pointed out that the sum of the deviations around the mean is zero, and accordingly the mean of the deviations around the mean is zero. Obviously, something must be done to alter this operation so that the mean deviation ($M.D.$) can reflect differences in variability.

Let us examine the reason why the sum of the deviations around the mean is always equal to zero ($\Sigma x = 0$). As shown in Chap. 5, the negative deviations are always exactly equal to the positive deviations, and they balance out to a sum of zero. Suppose the signs of the deviations were ignored and all deviations were considered positive. Then there would be no negative deviations to produce a sum of zero, and the mean deviation could be derived.

In computing the mean deviation the *absolute values* of the deviations from the mean are added, and the sum is divided by the number of cases. The absolute value of the deviations refers to the practice that all deviations are considered positive, whether they are actually positive or negative. The idea of absolute value is symbolized by placing vertical lines on each side of the variable that is always to be considered positive. The formula for the mean deviation is

$$\text{M.D.} = \frac{\Sigma |x|}{N}$$

where M.D. is the mean deviation

$|x|$ represents the deviations (disregarding signs) from the central value, in this case the mean

$N$ is the number of cases in the series.

**Illustration.** A group of older women differ in the number of children that they have had before the age of 30. Find the mean deviation for the distribution.

| X (number of children) | x |
|---|---|
| 1 | 1 |
| 4 | 2 |
| 3 | 1 |
| 1 | 1 |
| 0 | 2 |
| 2 | 0 |
| 2 | 0 |
| 1 | 1 |
| 5 | 3 |
| 1 | 1 |
| $\Sigma X = 20$ | $\Sigma |x| = 12$ |

$$\bar{X} = \frac{\Sigma X}{N} = \frac{20}{10} = 2$$

$$\text{M.D.} = \frac{\Sigma |x|}{N} = \frac{12}{10} = 1.2$$

**Problems. 1.** What is the mean deviation of the following scores received by a group of students on an attitude scale measuring social distance toward Negroes?

$$4, 6, 7, 11, 13, 4$$

*Ans.* 3

**2.** Find the mean deviation for the following distribution of grades in a course in human ecology:

$$82, 85, 73, 65, 84, \text{ and } 91$$

*Ans.* 7.33

## 6-8   Variation

Another measure of dispersion is based on squared deviations from the mean. The mean deviation uses absolute deviations around the mean, thereby preventing negative deviations and positive deviations from canceling one another and producing a sum of zero. Another way of avoiding negative signs is to square each deviation. To square a number, the number is multiplied by itself. For example, the square of 3 is 9; the square of 7 is 49. More important for our purpose here, the square of $-3$ is 9; the square of $-7$ is 49. The square of either a positive number or a negative number is always positive, since there would never be two unlike signs. (See the simplified explanation in Sec. 5-5.)

If we add together all the squared deviations from the mean, we have the *variation* around the mean. In our language, variation is $\Sigma x^2$. We discussed the sum of the squared deviations from the mean in the previous chapter (Sec. 5-5). Variation is just a new name for an old friend.

## 6-9   Variance

If the numbers in a distribution are about the same size and have approximately the same dispersion, then a group of 1,000 cases will certainly have a larger variation than a group of 50 cases. Adding squared deviations for twenty times as many cases will produce a larger sum. To offset this tendency, the number of cases is taken into account by dividing the variation by $N$, the number of cases. That will give us

$$\frac{\Sigma x^2}{N}$$

The name for this concept is *variance*, and it is symbolized by $\sigma^2$ (sigma squared). Variance is variation divided by the number of cases.

**Illustration.** Find the variance in submissiveness scores for a group of delinquent boys.

| $X$ | $x$ | $x^2$ |
|---|---|---|
| 13 | $-18.8$ | 353.44 |
| 22 | $-9.8$ | 96.04 |
| 41 | 9.2 | 84.64 |
| 18 | $-13.8$ | 190.44 |
| 37 | 5.2 | 27.04 |
| 22 | $-9.8$ | 96.04 |
| 19 | $-12.8$ | 163.84 |
| 16 | $-15.8$ | 249.64 |
| 35 | 3.2 | 10.24 |
| 74 | 42.2 | 1,780.84 |
| 61 | 29.2 | 852.64 |
| 19 | $-12.8$ | 163.84 |
| 40 | 8.2 | 67.24 |
| 29 | $-2.8$ | 7.84 |
| 31 | $-.8$ | .64 |
| $\Sigma X = 477$ | $\Sigma x = 0$ | $\Sigma x^2 = 4,144.40$ |

$$\bar{X} = \frac{\Sigma X}{N} = \frac{477}{15} = 31.8$$

$$\sigma^2 = \frac{\Sigma x^2}{N} = \frac{4,144.40}{15} = 276.29$$

## 6-10 Short Method for Calculating Variance

The illustration above indicates that finding the variance can be a time-consuming process if decimals are involved. Imagine how much worse the situation is when the mean has three decimals, for example, 31.762 rather than 31.8. All the deviations from the mean would have three decimal places, and the squared deviations would have six decimals. Therefore, we shall develop a method of finding the variance of a series without having to calculate the deviation of each case from the mean. All that is needed is knowledge of the rules of summation discussed in Chap. 4.

The variance is equal to the variation divided by the number of cases.

$$\sigma^2 = \frac{\Sigma x^2}{N}$$

For $x$, the deviation from the mean, let us substitute $(X - \bar{X})$. This substitution is correct because both expressions are different ways of indicating exactly the same thing. After all, the procedure in deriving the deviation is subtracting the mean from each case, and that is what $(X - \bar{X})$ signifies.

$$\sigma^2 = \frac{\Sigma(X - \bar{X})^2}{N}$$

To square a number, it is multiplied by itself. Let us perform the operation of squaring $(X - \bar{X})$ that is called for in the formula.

$$X - \bar{X}$$
$$X - \bar{X}$$

(Multiplying the first line by $- \bar{X}$) $\quad -\; X\bar{X} + \bar{X}^2$

(Multiplying the first line by $X$) $\quad X^2 -\; X\bar{X}$

(Adding) $\quad X^2 - 2X\bar{X} + \bar{X}^2$

Substituting $X^2 - 2X\bar{X} + \bar{X}^2$ for $(X - \bar{X})^2$, since we have just shown them to be equivalent, we obtain

$$\sigma^2 = \frac{\Sigma(X^2 - 2X\bar{X} + \bar{X}^2)}{N}$$

We can remove the parentheses in the numerator according to Rule 1 (Sec. 4-3).

$$\sigma^2 = \frac{\Sigma X^2 - \Sigma 2X\bar{X} + \Sigma \bar{X}^2}{N}$$

Both $\bar{X}$ and 2 are constants, not changing for this set of data. Of course, if $\bar{X}$ is a constant, then the square of it, $\bar{X}^2$, is also a constant. Let us move the two constants in the middle term to the left of the summation sign. According to Rule 2 (Sec. 4-4),

$$\sigma^2 = \frac{\Sigma X^2 - 2\bar{X}\Sigma X + \Sigma \bar{X}^2}{N}$$

The third term in the numerator, $\Sigma \bar{X}^2$, can be written $N\bar{X}^2$ according to Rule 3 (Sec. 4-5).

$$\sigma^2 = \frac{\Sigma X^2 - 2\bar{X}\Sigma X + N\bar{X}^2}{N}$$

Instead of dividing the entire numerator by $N$, we can do the same thing by dividing each term of the numerator by $N$. For the middle term, we can divide either $2$ or $\bar{X}$ or $\Sigma X$ by $N$, but we must be careful to divide by $N$ only once. Dividing any of the factors in that term by $N$ has the effect of dividing the result by $N$.

$$\sigma^2 = \frac{\Sigma X^2}{N} - 2\bar{X}\left(\frac{\Sigma X}{N}\right) + \frac{N\bar{X}^2}{N}$$

In the last term we multiply $\bar{X}^2$ by $N$ and also divide it by $N$. The two processes cancel each other, leaving only $\bar{X}^2$.

$$\sigma^2 = \frac{\Sigma X^2}{N} - 2\bar{X}\left(\frac{\Sigma X}{N}\right) + \bar{X}^2$$

The middle term contains the expression $(\Sigma X/N)$. This is the same as our definition of the mean (Sec. 5-3). We can insert $\bar{X}$ where $(\Sigma X/N)$ appears. That will give us $-2\bar{X}(\bar{X})$ in the middle term, which can be written less awkwardly as $-2\bar{X}^2$. Therefore,

$$\sigma^2 = \frac{\Sigma X^2}{N} - 2\bar{X}^2 + \bar{X}^2$$

An additional simplification is now possible. The second term indicates that the square of the mean should be subtracted twice, while the third term indicates that the square of the mean should be added. If we add something once and subtract it twice, we essentially are subtracting it only once. Accordingly the second and third terms can be combined by subtracting the squared mean once.

$$\sigma^2 = \frac{\Sigma X^2}{N} - \bar{X}^2$$

Replacing the mean by its definition, we finally arrive at

$$\sigma^2 = \frac{\Sigma X^2}{N} - \left(\frac{\Sigma X}{N}\right)^2$$

**Illustration.** Find the variance in submissiveness scores for the same group of delinquent boys.

| $X$ | $X^2$ |
|---|---|
| 13 | 169 |
| 22 | 484 |
| 41 | 1,681 |
| 18 | 324 |
| 37 | 1,369 |
| 22 | 484 |
| 19 | 361 |
| 16 | 256 |
| 35 | 1,225 |
| 74 | 5,476 |
| 61 | 3,721 |
| 19 | 361 |
| 40 | 1,600 |
| 29 | 841 |
| 31 | 961 |
| $\Sigma X = 477$ | $\Sigma X^2 = 19,313$ |

$$\sigma^2 = \frac{\Sigma X^2}{N} - \left(\frac{\Sigma X}{N}\right)^2$$

$$= \frac{19,313}{15} - \left(\frac{477}{15}\right)^2$$

$$= 1,287.5 - (31.8)^2$$

$$= 1,287.5 - 1,011.24$$

$$= 276.26$$

The result of 276.26 is almost the same (276.29) as that found by the longer method. The difference is the result of approximations introduced by rounding decimals. Both methods would give exactly the same result if the work were carried out to more decimal places.

This alternative formula for the variance saves time in computation although it may appear more difficult. The saving is the result of not having to square the deviation of each case from the mean, thereby avoiding numbers with many decimal places.

## 6-11  Standard Deviation

We shall now describe the most statistically important measure of dispersion. It is known as the standard deviation. Along with the mean, it is widely used in statistical analysis. Its stability and its ease of mathematical manipulation make it especially useful. Most of the techniques and procedures covered in the second part of this book dealing with statistical inference make extensive use of it. Therefore, it is especially important that this section be mastered thoroughly.

The *standard deviation*, which usually is symbolized by $\sigma$ (small Greek sigma), is the square root of the variance of a distribution. The square root of a number is found by finding another number that, multiplied by itself, will give the original number. If $\sigma^2$ is 64, then $\sigma$ is 8. Again, if the variance is 4, then the standard deviation is 2. If $\sigma$ is 5, then $\sigma^2$ is 25.

In order to clarify the relationship between the concepts variance and standard deviation, let us first of all reconsider the two variance formulas. The first, expressed as a definition, is

$$\sigma^2 = \frac{\Sigma x^2}{N}$$

The second formula, easier to use for computation, is

$$\sigma^2 = \frac{\Sigma X^2}{N} - \left(\frac{\Sigma X}{N}\right)^2$$

Since the standard deviation is the square root of the variance, we automatically have two formulas for the standard deviation. One merely finds the variance and then takes the square root of that number. Thus,

$$\sigma = \sqrt{\frac{\Sigma x^2}{N}}$$

$$\sigma = \sqrt{\frac{\Sigma X^2}{N} - \left(\frac{\Sigma X}{N}\right)^2}$$

Finding the square root of a number sometimes becomes troublesome for a great many students in elementary statistics. Table A in the Appendix contains a valuable aid—the square roots and squares of numbers. By utilizing this table you are taking advantage of the labor of other persons, thus saving time as well as relieving yourself of a tedious task. But you cannot use the table unless you know its theoretical basis. Therefore, it is essential to read carefully the instructions at the beginning of Table-A. It will be found a worthwhile investment, since you will save yourself hours of effort during the course.

**Problem.** Find the variance and standard deviation for the following data which give the number of siblings for each member of an experimental group. Do the problem by both the long and short methods.

$$0, 1, 0, 2, 1, 2, 4, 1, 2, 0, 3, 0$$

*Ans.* 1.556; 1.25

Definitions and names are fundamentally matters of choice. Some persons do not realize this and ask questions like, "Why is that called the variation and that the variance?" The term variance and its definition are examples of shared symbols. As long as we agree on the meaning of the term we can communicate effectively. This particular name is widely used, so we save effort by learning and using words and symbols that can be understood by scientists all over the world.

## 6-12   Coefficient of Variability or Coefficient of Variation

Various measures of dispersion have now been covered. Two applications of the most important measure, standard deviation, will be discussed. With a slight modification, the standard deviation can be used for comparing the relative dispersion of two or more groups which have different means and use different scales. We might like to know, for example, whether a class varies more on grades or on intelligence tests. Or we might want to determine whether there is more variability among Jews or among Negroes with respect to their attitude toward education.

The difficulty that precludes our using the standard deviation alone as a means of comparing two or more dispersions is that the size of the standard deviation may be influenced by the units and means of the distributions. Suppose we administered two tests to a single group. These tests are exactly alike, except that the scores on the first are from 1 to 10, and on the second from 10 to 100. It can be seen readily that the standard deviation of the second test is ten times as large as the standard deviation of the first, even though both series conform to the same pattern of dispersion. The variance of the second test is one hundred times as large as that of the first.

There is a way out of this problem. If the mean on the first test were 6, the mean on the identical second test would be 60, ten times as large. Therefore, the mean is affected by the units used in exactly the same way that the standard deviation is affected. The measure of variability we shall use for comparing dispersions will divide the standard deviation by the mean and thereby be uninfluenced by the units of analysis. It should not be used uncritically, however, since it may give misleading results.

The *coefficient of variability* (*V*) is defined as the standard deviation times 100 divided by the mean. Another name for the same thing is the *coefficient of variation*. Both names are in common use.

$$V = \frac{\sigma}{\bar{X}} (100)$$

It can be seen that the coefficient of variability is a measure of relative dispersion.

**Problem.** A men's physical-education class reported a mean height of 69 inches with a $\sigma$ of 4 inches and a mean weight of 160 pounds with a $\sigma$ of 20 pounds. Is the class more variable in height or in weight?

*Ans.* More variable in weight

## 6-13   Standard Score

The final measure studied in this chapter is not a measure of variability. It is used to determine the position of a single case in a distribution in such a manner as to make it possible to compare the positions of cases in different distributions. It is included at this point because it is based on the standard deviation.

If a student takes a test in each of two courses, getting 70 in one and 80 in the other, could we be sure that the 80 is actually a better grade? No, it is conceivable that the 70 was the highest grade in the class and the 80 the lowest grade in the other class.

Knowledge of the mean in each of the two classes of course would be helpful. If the mean on the first test were 58 and the mean on the second were 69, that would provide a basis for judging the student's relative standing on each test. He was 12 points above the mean on one test and 11 points above on the other. The relative importance of the 12 and 11 points depends, however, upon the variability of the other students in each class.

The *standard score* (*z*) expresses the score of an individual as a deviation from the mean in standard deviation units. The individual is thus placed accurately in the group, taking into account the group mean and the group standard deviation.

$$z = \frac{x}{\sigma} = \frac{(X - \bar{X})}{\sigma}$$

**Illustration.** Algernon gets 93 in history and brags to Arbuthnot, who in turn insists that his 81 in calculus is a more outstanding achievement.

|  | History | Calculus |
|---|---|---|
| $\bar{X}$ | 78 | 59 |
| $\sigma$ | 9 | 16 |

For Algernon,

$$z = \frac{x}{\sigma} = \frac{93 - 78}{9} = \frac{15}{9} = 1.67$$

For Arbuthnot,

$$z = \frac{x}{\sigma} = \frac{81 - 59}{16} = \frac{22}{16} = 1.38$$

*Ans.* Algernon can continue bragging.

**Problem.** A student gets a score of 19 in feeling of acceptance in a group and a score of 16 in friendliness toward that group. For feeling of acceptance, the mean is 31, and the standard deviation is 11. For friendliness, the mean is 35, and the standard deviation is 14. In which is he higher?

*Ans.* The $z$ score of $-1.09$ is higher than the $z$ score of $-1.36$. He is higher in feeling of acceptance.

## 6-14  Summary

The range is the difference between the largest and smallest value. It is easy to compute, but it has several disadvantages:

*a.* It is completely dependent on the two most extreme cases.

*b.* It is therefore an unstable measure.

*c.* As the number of cases increases, the range tends to increase.

The 10-90 percentile range is found by subtracting the 10th percentile from the 90th percentile. It is less unstable than the range. It is not affected as much as the range by the number of cases.

The three quartiles ($Q_1$, $Q_2$, and $Q_3$) divide the distribution into four parts (also called quartiles) with an equal number of cases in each part.

$Q$ (the semi-interquartile range) $= \dfrac{Q_3 - Q_1}{2}$.

M.D. (mean deviation) $= \dfrac{\Sigma|x|}{N}$.

Variation $= \Sigma x^2$.

$\sigma^2$ (variance) $= \dfrac{\Sigma x^2}{N} = \dfrac{\Sigma X^2}{N} - \left(\dfrac{\Sigma X}{N}\right)^2$.

$\sigma$ (standard deviation) $= \sqrt{\dfrac{\Sigma x^2}{N}} = \sqrt{\dfrac{\Sigma X^2}{N} - \left(\dfrac{\Sigma X}{N}\right)^2}$.

The standard deviation is the most widely used measure of dispersion.

$V$ (coefficient of variability) $= \dfrac{\sigma}{\bar{X}} (100)$.

$z$ (standard score) $= \dfrac{x}{\sigma}$.

## PROBLEMS FOR CHAP. 6

1. A group of mothers are rated by means of a questionnaire on their adjustment during pregnancy. This information will be used to relate pregnancy adjustment of the mother to the adjustment of the child in the first 2 years of life. For 20 mothers, the pregnancy-adjustment scores were 18, 22, 27, 30, 31, 32, 33, 35, 36, 39, 40, 41, 41, 41, 42, 44, 44, 47, 57, and 60.

   a. What is the range in pregnancy adjustment?   *Ans.* 18 to 60, or 42
   b. What is the 10-90 percentile range?      *Ans.* 24.5 to 52, or 27.5
   c. What is $Q$?                                              *Ans.* 5.75
   d. What is the mean?                                          *Ans.* 38
   e. What is the mean deviation?                              *Ans.* 7.8
   f. What is the variation?                                 *Ans.* 2,010
   g. What is the variance?                                  *Ans.* 100.5
   h. What is the standard deviation?                        *Ans.* 10.0
   i. What is $V$?                                            *Ans.* 26.3
   j. Express a score of 31 as a standard score.            *Ans.* $-.7$
   k. What is the $z$ score equivalent to 60?                 *Ans.* 2.2

# 7 Ratios, Proportions, and Rates

The preceding chapters on measures of central value and dispersion are basic to descriptive statistics. In this chapter we deal with several other summarizing measures, those that relate to the number of cases that have, or do not have, some particular characteristic.

## 7-1  Quantitative and Qualitative Variables

In attempting to describe a human being, it is virtually impossible to list every detail about him. No matter how many characteristics are indicated, such as age, height, foot size, and general physical appearance, there is always a multiplicity of other characteristics which are omitted. Generally, only those characteristics that are significant and pertinent with respect to a particular problem are the ones selected for consideration. The automobile salesman may be interested in your income, but, in all probability, not in your usual breakfast menu.

Some characteristics are described as *quantitative*, referring to variables that are measured numerically. In compiling a list of characteristics of an adult male, the following are examples of quantitative variables: height, 64 inches; weight, 312 pounds; IQ, 36; score on an honesty test, 189.

Other characteristics may refer to *qualitative* or nonquantitative variables. Qualitative variables operate on an all-or-nothing basis. Categories are set up in advance. Each individual case is then assigned to a category or excluded from it. There is no possibility of being only half in. For example, an adult male may be described in terms of the following qualitative characteristics: short, overweight, feeble-minded, and honest. In themselves, these statements mean very little, but when the definition of each category is given, the statements take on more significance. For example, we may specify in advance that any height below 66 inches will be referred to as "short," that heights between 66 inches and 71 inches will be labeled "average," and those over 71 inches "tall." For each case, the only problem is to determine the appropriate category.

Sometimes people attempt to draw a sharp line between quantitative and qualitative characteristics, acting as if characteristics were inherently one or the other. This view is not correct. The qualitative or quantitative nature of a characteristic is the result of a decision by the investigator. No doubt it was observed that the same characteristics of the person were discussed in both the preceding paragraphs. In one paragraph the characteristics are quantitative; in the other, qualitative.

Every quantitative variable can be changed into a qualitative variable. For example, cases falling between two points on a scale may be placed in one category, and all other cases in another category. Often a name is given to each category. Suppose a group of students disclose on a questionnaire the number of times they had cheated on examinations during the preceding school year. The facts indicated on the questionnaires would be expressed in the form of a quantitative variable, ranging from 0 to some fairly large number. These data could be easily transformed into a qualitative variable. Persons indicating 0 on the questionnaire could be placed in one class, those indicating 1 or 2 in a second class, and the remainder in a third category. If desired, each of the three classes could be given a label, such as "noncheaters," "occasional cheaters," and "frequent cheaters." The boundaries of the classes and their names are completely arbitrary, depending on the researcher's purpose.

It also is possible to transform qualitative statements into quantitative ones. In physics, qualitative colors are quantified on the basis of wavelengths. We often have difficulty in developing appropriate units for quantitative analysis in the social sciences, but a good start has been made in that direction. For example, instead of describing a person as bigoted, he is characterized as having a high score on a scale measuring religious prejudice. For income-tax purposes the Bureau of Internal Revenue allows an extra deduction for the blind, but it carefully defines blindness in quantitative terms. A person is blind if his corrected vision in both eyes is worse than 20/200 or if his vision does not cover more than a 20-degree angle.

Even if no basic units for quantitative analysis are available, qualitative statements are often changed into quantitative ones. This is done by arbitrarily assigning numerical weights to the categories. For example, a group of people might be classified into conventional qualitative political categories—Democrats, Independents, and Republicans. If each Democrat is given a score of 0, each Independent a score of 1, and each Republican a value of 2, then one can obtain a quantitative measure of the degree of "Republicanness" of a group. In applying such arbitrary assignments of weights, it should be recognized that the final results are only as good as the appropriateness and validity of the weights that are chosen. Inappropriate application of weighting procedures can give absurd results. Experience and insight are the best insurance against pitfalls of this kind.

## 7-2  Ratios

This chapter is concerned with three measures that are used for variables that appear in a qualitative form. They are used in descriptive statistics to give additional information that goes beyond the number of persons who fall in a class.

The first and most general measure is the *ratio*. It shows the relation in size between two numbers. If a group of 30 students consists of 12 boys and 18 girls, the ratio of boys to girls is 12:18. This ratio also can be written as 12/18 or, dividing the numerator and denominator by 6, as 2/3. Similarly, the ratio of girls to boys is 18:12, 18/12, 3/2, or 1.5.

**Problem.** A group is composed of 67 Caucasoids, 19 Negroids, and 5 Mongoloids. What is the ratio of Caucasoids to Negroids? Of Negroids to Caucasoids? Of Mongoloids to Negroids? Of Mongoloids to Caucasoids? Of Negroids to Mongoloids? Of Caucasoids to Mongoloids?

*Ans.* 67/19; 19/67; 5/19; 5/67; 19/5; 67/5

Some ratios are used so frequently that the basis is not stated, since it is assumed that the reader is familiar with them. For example, in social research, the *sex ratio* is very widely used. It indicates number of males for every 100 females. If the sex ratio for some community is 75, that means that there are 75 males per 100 females, or 3 males for every 4 females.

## 7-3   Proportions

A *proportion* is a special type of ratio in which the denominator is the entire group. For example in the preceding problem (Sec. 7-2) there are 67 Caucasoids, 19 Negroids, and 5 Mongoloids. The total number of persons in the group is

$$67 + 19 + 5 = 91$$

The proportion of Caucasoids is 67/91, the proportion of Negroids is 19/91, and the proportion of Mongoloids is 5/91.

Notice that the sum of the three proportions above is exactly 1:

$$\frac{67}{91} + \frac{19}{91} + \frac{5}{91} = \frac{91}{91} = 1$$

If a group is divided into several classes and the proportion in each group is determined, then the sum of all the proportions must be 1. If there are only two categories, then the sum of the two proportions must be 1.

**Proof.** If a group is divided into two parts, then the proportion in the first group plus the proportion in the second group must equal 1.

Let $p$ = the proportion in the first group

$q$ = the proportion in the second group

$A$ = the number in the first group

$B$ = the number in the second group

$N$ = the total number in the entire group

$p = \dfrac{A}{N}$, by definition of a proportion

$q = \dfrac{B}{N}$, by definition of a proportion

By adding the two proportions, we have the expression

$$\frac{A}{N} + \frac{B}{N}$$

This can be written as

$$\frac{A + B}{N}$$

But adding $A$, the number in the first group, to $B$, the number in the second group, must give $N$, the number in the whole group. We can, therefore, replace $A + B$ by $N$. Thus

$$\frac{A + B}{N} = \frac{N}{N} = 1$$

**Problem.** A group of 43 St. Louis clubwomen were asked where they resided a year previously. Their responses showed that 36 were living at the same address, 4 had moved within the city, and 3 had moved from suburban areas.

What proportion of the clubwomen

| | |
|---|---|
| *a.* Had moved within the city? | *Ans.* 4/43 |
| *b.* Had not moved at all? | *Ans.* 36/43 |
| *c.* Had moved from the suburbs? | *Ans.* 3/43 |
| *d.* Had moved? | *Ans.* 7/43 |

What proportion of those moving

| | |
|---|---|
| *e.* Had moved within the city? | *Ans.* 4/7 |
| *f.* Had moved from suburban areas? | *Ans.* 3/7 |

## 7-4   Rates

A *rate* is a special type of ratio in which the numerator is the number of times a specified kind of event occurs during a particular time period and the denominator ideally is the whole number of exposures to the risk of its occurrence. The validity of a rate depends upon a proper choice of both numerator and denominator. The derived ratio is usually multiplied by some large number, like 1,000, in order to express rates as whole numbers rather than as decimals. A few illustrations should make clear the concept of rate, with the denominator representing an approximation to the number of exposures.

The formula for the crude birth rate is

$$\text{Crude birth rate} = \frac{\text{number of live births in an area during year}}{\text{population of area at midyear (July 1)}} \times 1{,}000$$

If there were 80,000 live births in a state having a population on July 1 of 8,000,000, then

$$
\begin{aligned}
\text{Crude birth rate} &= \frac{80{,}000}{8{,}000{,}000} \times 1{,}000 \\
&= \frac{80{,}000{,}000}{8{,}000{,}000} \\
&= 10.0
\end{aligned}
$$

The crude birth rate, like other crude rates, is based on the total population and represents only a very approximate measure. In order to "refine" the crude birth rate, all members of the male sex and all single, widowed, and divorced females, as well as married women above the childbearing age, should be eliminated from the denominator of the formula, since they normally would not be "exposed" to the condition which is being measured.

Therefore, a more precise rate would be the nuptial birth rate, which is computed according to the following formula:

$$\text{Nuptial birth rate} = \frac{\substack{\text{number of live legitimate births} \\ \text{in an area during the year}}}{\substack{\text{number of married females, 15} \\ \text{to 44 years of age}}} \times 1{,}000$$

The formula for the crude death rate is

$$\text{Crude death rate} = \frac{\substack{\text{number of deaths in an area dur-} \\ \text{ing a given year}}}{\text{population of area at midyear}} \times 1{,}000$$

Similarly, a crude crime rate, juvenile-delinquency rate, suicide rate, or a crude rate of some other social phenomenon may be indicated in accordance with the same basic formula.

A more refined measure of mortality, for example, is the specific death rate. The general equation is

$$\text{Specific death rate} = \frac{\text{deaths in a specified class or group in a year}}{\text{midyear population in same specified class or group}} \times 1{,}000$$

An age-sex specific death rate, for example, would measure mortality for some group, such as males between 60 to 64 years of age. A series of rates of this kind would be more reliable and more meaningful than a crude rate.

## 7-5    Difficulties in Deriving Appropriate Rates

The population at the middle of the year (July 1) is used above as the most representative estimate of the number of persons in the population during the year. In any relatively large area the population is undergoing constant change. If, for example, it is showing a steady increase, the population at the beginning of the year would not be a suitable denominator for computing a rate, since it is at its lowest point. On the other hand, the population at the end of the year would not be satisfactory, since it would be higher on that date than at any point during the preceding 12 months. The midyear population is used as a substitute for the mean, which would theoretically be the most satisfactory measure. In order to calculate the mean, however, we would have to know the exact population for every day during the year. Therefore, the midyear population is used as an approximation to the number of exposed persons.

It cannot be overemphasized that in computing rates it is important that the denominator and numerator bear a direct relationship to each other. This often demands laborious, yet necessary, effort to eliminate extraneous events. For example, in the crude birth rate we are interested in all births that occur in the population that is resident in a particular area. That means that births that occur to nonresidents in the area must be eliminated. Correspondingly, births to residents temporarily away from an area are added to the total. The National Office of Vital Statistics has the responsibility for tabulating and properly allocating birth data for the entire country. Their reports provide statistics on births by both place of occurrence and by place of residence. Similar data are also available for deaths.

It is essential to understand the logical basis as well as the difficulties involved in deriving appropriate and meaningful rates. Such understanding will help in developing ingenuity in setting up rates, as well as in providing a healthy skepticism when reading reports that use ambiguous or erroneous rates.

Let us attempt to develop a procedure for computing a series of rates by census tracts for an ecological study of juvenile delinquency. This problem will reveal some of the difficulties that the social scientist must face in devising appropriate and reliable measures of the incidence of juvenile delinquency and of other phenomena.

The first problem is to procure a measure or index of juvenile delinquency for each of the census tracts. The most generally available, as well as the most widely used, index is the cases that come to the official attention of the juvenile court and the police department. Data of this kind never represent a complete record of all cases that occur. For one reason or another a considerable number of cases is excluded because (1) many juvenile offenders are never apprehended; (2) sometimes known delinquents may not be reported; (3) known violators are frequently handled unofficially by psychiatrists, social agencies, and boarding schools.

Moreover, the proportion of cases actually reported may vary considerably from one census tract to another. Such extraneous factors as prevailing attitudes and traditions may exert a strong influence on the reporting of delinquencies. For example, in some areas the attitudes toward infractions by juveniles may be very tolerant, while in other areas they may be very severe.

It should be recognized that the unit for statistical analysis of juvenile delinquency as normally defined also possesses serious limitations from a qualitative point of view. It is nonspecific in character, including everything from truancy, malicious mischief, incorrigibility, and traffic violations to larceny, burglary, robbery, and even homicide.

The next problem in deriving rates is to allocate the cases according to census tracts. The residence of the delinquent rather than the place where the delinquency was committed is the most commonly used criterion. If "place of occurrence" rather than "residence" were used as a basis for allocating cases, it would be extremely difficult, if not impossible, to relate them to any logical population factor.

Moreover, crude juvenile-delinquency rates would be entirely unsatisfactory, since the number of children, delinquent and nondelinquent, defined by statute as juvenile, varies markedly from one part of the city to another. In other words, it is essential to know exactly the number of persons who are "exposed" to the risk of committing juvenile delinquency in each of the census tracts. Since the incidence of delinquency for boys is so very much higher than it is for girls, rates should be computed separately. One formula that we could use is

$$\text{Juvenile-delinquency rate for boys} = \frac{\text{number of male juveniles, as defined by statute, residing in a specified area, committing delinquencies during a given period}}{\text{male population of juvenile court age living in specified area at the mid-point of the given period}} \times 1{,}000$$

In order to ensure an adequate sample, and hence greater reliability for our rates, a 2-, 3-, or even 4-year period may be covered. The population used still would represent the mid-point of the period. To adjust to an annual base the total rate, of course, would be divided by the number of years in the period covered.

## 7-6   Graphic Portrayal of Rates

Rates may be computed for many different types of comparisons. The accompanying figures graphically portray three types of

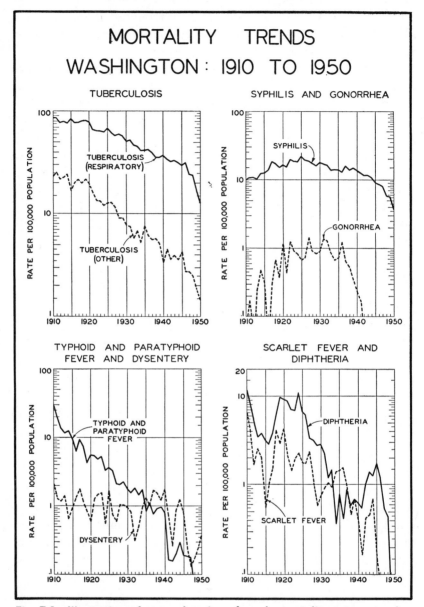

Fig. 7-1  Illustration of several series of crude mortality rates covering trends for eight specified causes of death for a 40-year period. The curves are plotted on a semilogarithmic type of grid. See Chap. 3 for a more detailed discussion of this particular graphic form.

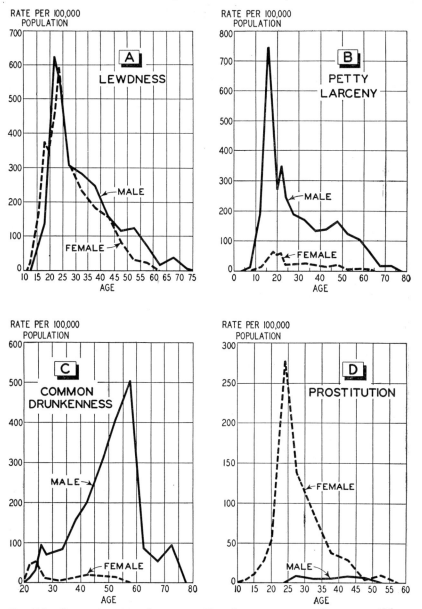

**Fig. 7-2   Several series of age-specific crime rates according to sex.** These data are for the city of Seattle for the 2-year period 1950–1951.

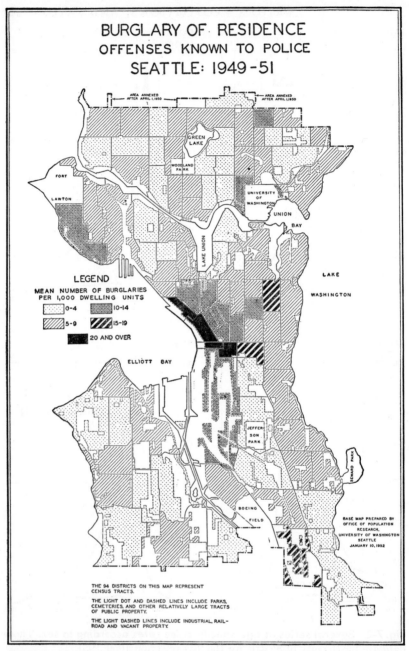

**Fig. 7-3 Map portraying series of crime rates by census tracts for a large city.** It will be observed that the crime shown on this map represents cases of burglary of residence which have been reported to the Seattle Police Department during the 3-year period 1949–1951. The rates are based on dwelling units rather than population.

rates frequently found in social research. Figure 7-1 shows trends in mortality according to certain specified causes of death from 1910 to 1950 in the state of Washington. It will be observed that the crude mortality rates per 100,000 of total population for each of the eight diseases indicated on the chart have declined considerably in recent years.

The crime rates presented in Fig. 7-2 depict crime rates for specific age groupings for each of the two sexes. Even a casual inspection of the graphs provides a basis for clear differentiation and comparison according to both age and sex. Many more males than females are arrested for petty larceny and common drunkenness, while most arrestees for prostitution are female. There is about an equal proportion of males and females arrested for lewdness. Except for common drunkenness, most offenders are young, with comparatively few arrested after the age of 30.

The spatial distribution of one of the major crimes is presented according to census tracts for a large city in Fig. 7-3. The incidence of burglary of residence is based on dwelling units rather than on population. The area in and near the central portions of the city has by far the highest burglary rate. Such information as shown in this chart possesses both utilitarian and scientific value. It can be used for administrative purposes by social-welfare agencies, the police and other law-enforcing agencies, and by sociologists in testing scientific hypotheses.

## 7-7  Summary

Quantitative variables can be changed into qualitative variables, and qualitative variables can be changed into quantitative variables.

Ratios show the relation in size between two numbers.

In a proportion $(p,q)$, the denominator is the size of the whole group.

In a rate, the numerator is the number of events which occur during some specific time period, and the denominator is directly related to the numerator.

**PROBLEMS FOR CHAP. 7**

1. In a predominantly Negro area of Chicago, 9,347 nonwhite females reported their occupations out of 9,527 employed nonwhite females. The occupational distribution is as follows:

| Occupation | Number |
|---|---|
| Total | 9,347 |
| Operatives | 3,650 |
| Other service workers | 2,070 |
| Private household workers | 1,551 |
| Clerical workers | 737 |
| Laborers | 468 |
| Professional workers | 266 |
| Salesworkers | 253 |
| Craftsmen and foremen | 214 |
| Managers and officials | 138 |

*a.* What proportion is professional workers?            *Ans.* .03
*b.* What proportion is clerical or salesworkers?          *Ans.* .11
*c.* What is the ratio of laborers to operatives?          *Ans.* .13
*d.* What is the ratio of private household workers to salesworkers?
                                                         *Ans.* 6.13

2. In the year 1950, there were 2,593 births in a population which was 78,145 on January 1 and 80,005 on December 31. What is the crude birth rate?            *Ans.* 32.8

# 8 Computations with Grouped Data

·

Previous chapters emphasized basic ideas, not practical computation problems. This chapter introduces no new descriptive measures. It merely seeks to explain the principles by which the usual measures, such as the mean, median, and standard deviation, can be computed when the data are presented in grouped form. Some of these principles lead to the development of short cuts for computation.

## 8-1 Computation of Mean from Grouped Data

Let us continue to use for illustrative purposes the data in Table 2-3 on the age distribution of male forgers. Consideration will be given first to the computation of the mean.

If we assume that the mid-point of each class is the mean of the values for that class, then $\Sigma fm$ can replace $\Sigma X$ for that class. The cases in each class are replaced by the value of the mid-point of the class. The multiplication of the value of the mid-point by the number of cases in the class interval ($fm$) results in an identical product when the mid-point is equal to the mean of the cases in the interval. The formula for computing the mean from ungrouped data $\bar{X} = \Sigma X/N$ becomes $\bar{X} = \Sigma fm/N$ for grouped data.

105

**Table 8-1   Calculation of Mean by Long Method** (Data from Table 2-3, Age Distribution of Arrestees for Forgery)

| (1) | (2) | (3) | (4) |
|---|---|---|---|
| Class-interval age, years | Mid-point of class | Frequency | Frequency multiplied by mid-point of class |
| | $(m)$ | $(f)$ | $(fm)$ |
| 15–19 | 17.5 | 8 | 140.0 |
| 20–24 | 22.5 | 21 | 472.5 |
| 25–29 | 27.5 | 22 | 605.0 |
| 30–34 | 32.5 | 15 | 487.5 |
| 35–39 | 37.5 | 13 | 487.5 |
| 40–44 | 42.5 | 11 | 467.5 |
| 45–49 | 47.5 | 3 | 142.5 |
| 50–54 | 52.5 | 4 | 210.0 |
| 55–59 | 57.5 | 1 | 57.5 |
| 60–64 | 62.5 | 1 | 62.5 |
| | | $N = 99$ | $\Sigma fm = 3{,}132.5$ |

$$\bar{X} = \frac{\Sigma fm}{N}$$
$$= \frac{3{,}132.5}{99}$$
$$= 31.6$$

## 8-2   Short Method for Computing Mean

The computations in Sec. 8-1 are fairly laborious, although probably a little quicker than working with ungrouped data. We shall next develop a method that is much faster, noting first the principles upon which it is based.

The sum of the deviations around the mean is equal to zero (Sec. 5-4). The method we shall discuss starts by guessing the mean. The *guessed mean* $(\bar{X}')$ is a point chosen for convenience of tabulation and is unlikely to be the exact true mean. Therefore, the sum of the deviations around the guessed mean probably will not be equal to zero. Luckily, the sum of these deviations can be used as a correction factor to tell us the value of the true mean.

If the guessed mean is higher than the true mean, there will be too many cases below the guessed mean, and the sum of the deviations around the guessed mean will be negative. If the guessed mean is too low, then the sum of the deviations will be positive. By adding the mean amount of error to the guessed mean, we can obtain the true mean.

**Illustration.** The true mean of the following numbers is 5:

$$1, 4, 5, 7, 8$$

Let us suppose the mean is 4. The deviations from this assumed mean will be $-3$, 0, 1, 3, and 4. The sum of the deviations is $+5$. The mean amount of error is $5/N = 5/5 = 1$.   By adding 1 to 4, the result is 5.

Again, if the mean were assumed to be 6.5, the deviations would be $-5.5$, $-2.5$, $-1.5$, .5, and 1.5. The sum of the deviations is $-7.5$, making the correction factor $-7.5/5 = -1.5$. Adding $-1.5$ to 6.5 gives 5, the true mean.

One further principle will be used in the short-cut formula for computing the mean from grouped data. In finding the sum of the deviations around the guessed mean, it would save effort if the deviations were in convenient simple numbers. In grouped data, the deviation of a mid-point from the guessed mean would be $m - \bar{X}'$. If the deviation of each mid-point from the guessed mean is divided by the size of a class interval ($i$), the result is called the *step deviation* ($d'$).

$$d' = \frac{m - \bar{X}'}{i}$$

But dividing each deviation by the size of the class interval would change the numerical value of the result if no counteracting measure were taken. Therefore, after the sum of the deviations around the guessed mean is found in step-deviation units, that sum is multiplied by the size of the class interval. This restores the result to the original units.

The short formula for computing the mean from grouped data can now be developed. The true mean equals the guessed mean plus its correction factor.

$$\bar{X} = \bar{X}' + \frac{\Sigma f(m - \bar{X}')}{N}$$

The correction factor above is the mean of the sum of the deviations around the guessed mean. The frequencies in each class are assumed to be at the mid-point and have the mid-point's deviation from the guessed mean.

In expressing the correction factor in step-deviation units, it becomes necessary to divide by the size of the class interval. Therefore, we also must multiply by the size of the interval.

$$\bar{X} = \bar{X}' + \frac{\Sigma fd'}{N} \quad (i)$$

The formula we have developed is most easily applied to grouped data in which the size of the class interval remains the same for every class. This is why we stated that calculations are

Table 8-2    Calculation of Mean by Short Method (Data from Table 2-3, Age Distribution of Arrestees for Forgery)

| (1) | (2) | (3) | (4) | (5) |
|---|---|---|---|---|
| Class-interval age, years | Mid-point of class | Frequency | Deviations from guessed mean in step-deviation units | Frequency multiplied by deviation |
| | $(m)$ | $(f)$ | $(d')$ | $(fd')$ |
| 15–19 | 17.5 | 8 | −4 | −32 |
| 20–24 | 22.5 | 21 | −3 | −63 |
| 25–29 | 27.5 | 22 | −2 | −44 |
| 30–34 | 32.5 | 15 | −1 | −15 |
| 35–39 | 37.5 | 13 | 0 | 0 |
| 40–44 | 42.5 | 11 | 1 | 11 |
| 45–49 | 47.5 | 3· | 2 | 6 |
| 50–54 | 52.5 | 4 | 3 | 12 |
| 55–59 | 57.5 | 1 | 4 | 4 |
| 60–64 | 62.5 | 1 | 5 | 5 |
| | | $N = 99$ | | $\Sigma fd' = -116$ |

$$\bar{X} = \bar{X}' + \frac{\Sigma fd'}{N} \quad (i)$$
$$= 37.5 + \frac{-116}{99} \quad (5)$$
$$= 37.5 - 5.9$$
$$= 31.6 \quad \text{(see Fig. 8-1)}$$

GRAPHIC PORTRAYAL OF MEAN AND MEDIAN
AGE DISTRIBUTION OF ARRESTEES FOR FORGERY

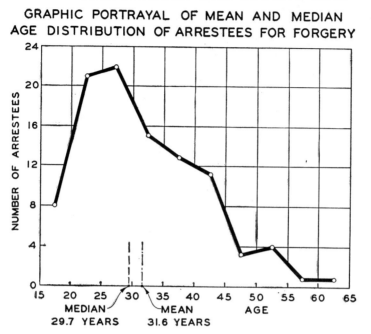

Fig. 8-1    Locating mean and median on a frequency polygon.

made easier if all class intervals are of equal size (Sec. 2-15). Although the formula may appear complicated, it is unquestionably the best method for computing the mean for grouped data with a large number of cases.

**Illustration.** Using the data from Table 2-3, let us assume that the mean is located at the mid-point of one of the central classes, 35 to 39. Any interval chosen will give the correct answer, but in order to keep the size of the numbers in the computation process to a minimum, care should be taken to select a class interval as close as possible to the one that contains the true mean, so far as one can judge.

The answer by this method should be exactly the same as the method used in Sec. 8-1. The only assumption made is the use of the mid-point of the class as representative of all the cases in that class, and this assumption is made in both the long and short methods. Both methods, therefore, are slightly inaccurate, but the saving in time outweighs the loss in accuracy.

## 8-3   Short Formula for Computing Standard Deviation

The short formula for finding the standard deviation from grouped data is derived in exactly the same manner. The long formula for grouped data is

$$\sigma = \sqrt{\frac{\Sigma f x^2}{N}}$$

**Table 8-3   Calculation of Standard Deviation by Short Method** (Data from Table 2-3, Age Distribution of Arrestees for Forgery)

| (1) | (2) | (3) | (4) | (5) | (6) |
|---|---|---|---|---|---|
| Class-interval age, years | Mid-point of class | Frequency | Deviations from guessed mean in step-deviation units | Frequency multiplied by deviation | Frequency multiplied by deviation squared |
| | $(m)$ | $(f)$ | $(d')$ | $(fd')$ | $(f)(d')^2$ |
| 15–19 | 17.5 | 8 | $-4$ | $-32$ | 128 |
| 20–24 | 22.5 | 21 | $-3$ | $-63$ | 189 |
| 25–29 | 27.5 | 22 | $-2$ | $-44$ | 88 |
| 30–34 | 32.5 | 15 | 1 | $-15$ | 15 |
| 35–39 | 37.5 | 13 | 0 | 0 | 0 |
| 40–44 | 42.5 | 11 | 1 | 11 | 11 |
| 45–49 | 47.5 | 3 | 2 | 6 | 12 |
| 50–54 | 52.5 | 4 | 3 | 12 | 36 |
| 55–59 | 57.5 | 1 | 4 | 4 | 16 |
| 60–64 | 62.5 | 1 | 5 | 5 | 25 |
| | | $N = 99$ | | $\Sigma fd' = -116$ | $\Sigma f(d')^2 = 520$ |

$$\sigma = i \sqrt{\frac{\Sigma f(d')^2 - \frac{(\Sigma fd')^2}{N}}{N}}$$

$$= 5 \sqrt{\frac{520 - \frac{(-116)^2}{99}}{99}}$$

$$= 5 \sqrt{\frac{520 - 136}{99}}$$

$$= 5 \sqrt{3.878}$$
$$= 5(1.97)$$
$$= 9.85 \quad \text{(see Figure 8-2)}$$

GRAPHIC PORTRAYAL OF STANDARD DEVIATION
AGE DISTRIBUTION OF ARRESTEES FOR FORGERY

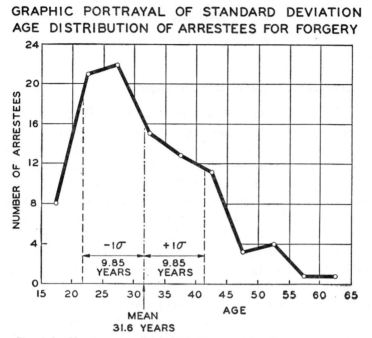

Fig. 8-2   Plotting standard deviation on a frequency polygon.

The deviations are again expressed in step-deviation units. Since the standard deviation is concerned with squared deviations, the deviations inside the square-root sign must be divided by $(i)^2$. Dividing by $(i)^2$ inside a square root is the same as dividing by $(i)$ outside the radical sign. Therefore, to restore the standard deviation to the original units, the short formula is multiplied by $(i)$ in front of the radical sign.

The correction factor for the standard deviation is always subtracted. The reason is that the sum of the squared deviations around the guessed mean must always be larger than the sum of the squared deviations around the true mean (Sec. 5-5).

Finally, because we are dealing with squared deviations, the correction factor for standard deviation must be the mean of the *square* of the sum of the step deviations around the guessed mean.

In the light of the three statements listed above, the short formula for finding the standard deviation is

$$\sigma = i \sqrt{\frac{\Sigma f(d')^2 - \dfrac{(\Sigma fd')^2}{N}}{N}}$$

## 8-4 Computing Median from Grouped Data

Our final task is to develop a method for finding the median when the data are presented in group form. We shall reverse our usual procedure and find the median before we state the formula. The reason for this reversal is that the formula merely restates a common-sense method in a rather frightening form.

Once again we shall turn to Table 2-3 for our illustrative data. The median in a grouped distribution is that point which divides the cases so that half are below the median and half are above. Since there are 99 cases, the median should have 49.5 cases below it. Although the cases are discrete, we shall treat them as continuous (Sec. 2-6).

At the lower limit of the second class, the cumulative frequency is 8, the number in the first class. At 25, the lower limit of the next class, 29 cases have been passed. When we reach 30, the lower limit of the fourth class, the number of cases below that point has reached 51. We want the point that has exactly 49.5 cases below it. That point must be in the class whose lower limit is 25, with 29 cases below it, and whose upper limit is almost 30, with 51 cases below.

Having determined which class contains the median, the next step is to determine how far into that class one must go in order to find the median. This is known as interpolation. Before entering the class, 29 cases have been accounted for. Then 20.5 more cases must be passed before the median will have 49.5 cases below it.

At this point we introduce a new assumption. Instead of considering every case as being at the mid-point of the class, let us assume that the cases are evenly spread throughout the class. We need some description of the manner in which the cases are spread through the class interval, and lacking any real knowledge, this assumption is reasonable.

If we need 20.5 additional cases and there are 22 cases evenly spread through the interval, the median must be relatively far into that interval. We go into the interval exactly (20.5/22) of the entire distance. What is the interval? It is almost 5, so we move into that class (20.5/22)(5) = 4.7 units. That is, 4.7 units

Table 8-4  Calculation of the Median (Data from Table 2-3, Age Distribution of Arrestees for Forgery)

| (1) | (2) | (3) |
|---|---|---|
| Class-interval age, years | Frequency | Cumulative frequency |
| | $(f)$ | |
| 15–19 | 8 | 8 |
| 20–24 | 21 | 29 |
| 25–29 | 22 | 51 |
| 30–34 | 15 | 66 |
| 35–39 | 13 | 79 |
| 40–44 | 11 | 90 |
| 45–49 | 3 | 93 |
| 50–54 | 4 | 97 |
| 55–59 | 1 | 98 |
| 60–64 | 1 | 99 |
| | $N = 99$ | |

$$\text{Median} = l + \frac{N/2 - F}{f} \quad (i)$$

$$= 25 + \frac{99/2 - 29}{22} \quad (5)$$

$$= 25 + \frac{20.5}{22} \quad (5)$$

$$= 25 + 4.7$$

$$= 29.7 \quad \text{(see Figure 8-1)}$$

of that interval must be added to the lower limit of the interval, at which point we already had 29 cases below. The lower limit is 25, which, when added to 4.7, gives the median as 29.7.

The formula for finding the median from grouped data forces us to go through this same process in a mechanical fashion. Having determined the class in which the median must be placed, substitutions are made in the formula.

$$\text{Median} = l + \frac{N/2 - F}{f} \ (i)$$

The two new terms are $l$ and $F$. The $l$ stands for the lower limit of the class containing the median, and the $F$ is the cumulative frequency up to that lower limit. The $f$ is the frequency of the class containing the median, and $i$ is the size of that class interval.

Table 8-4 illustrates the step-by-step procedure in computing the median.

## 8-5  Computing Mode from Grouped Data

It will be recalled from our discussion of ungrouped data that the mode is the most frequently occurring value (Sec. 5-9). There are several methods for determining the mode in a frequency distribution, but none is universally accepted. The simplest method, satisfactory for most practical purposes, is based on inspection of the frequency distribution. The modal class, the interval with the largest frequency, is ascertained, and the mid-point is considered the mode. In Table 8-2, the mode is 27.5.

## 8-6  Summary

$$\bar{X} = \frac{\Sigma fm}{N}$$

$$\bar{X} = \bar{X}' + \frac{\Sigma fd'}{N} \ (i)$$

$$\sigma = \sqrt{\frac{\Sigma fx^2}{N}}$$

$$\sigma = i \sqrt{\frac{\Sigma f(d')^2 - \frac{(\Sigma fd')^2}{N}}{N}}$$

$$\text{Median} = l + \frac{N/2 - F}{f} \ (i)$$

## PROBLEMS FOR CHAP. 8

1. A group of boys at a summer camp were asked to make socio-metric choices for eight different activities. The popularity of each

boy was measured by the total number of choices received by him. The data, summarized in tabular form, are presented below.

| Number of choices received ($X$) | Number of boys ($f$) |
|---|---|
| Total | 91 |
| 2–3 | 1 |
| 4–5 | 3 |
| 6–7 | 14 |
| 8–9 | 22 |
| 10–11 | 19 |
| 12–13 | 17 |
| 14–15 | 9 |
| 16–17 | 4 |
| 18–19 | 0 |
| 20–21 | 1 |
| 22–23 | 1 |

a. What is the median number of choices received?     *Ans.* 10.1
b. What is the mean?     *Ans.* 10.4
c. What is the standard deviation?     *Ans.* 3.45

# 9  Universe and Sample

sample
                     population, or universe
                     statistic
                     parameter


We now have completed a survey of the most frequently used measures of central tendency and dispersion. In the remainder of the book these descriptive measures will be used in a new context, commonly referred to as statistical inference or inductive statistics. This chapter serves as a general introduction to statistical inference, explaining the setting in which generalizations are made from sample results.

## 9-1  Sample and Population

If we were interested in the mean age of entering freshmen at a large state university, our computations might be based on a tabulation of every freshman in the institution. Such a procedure would be extremely laborious and time-consuming. Instead, it would be much easier and, for all practical purposes, just as accurate to select a sample of freshmen for analysis from which to draw conclusions. Most statistical studies are based on samples and not on complete enumerations of all relevant data. A statistical *sample* represents only part of an entire group. The entire group from which a sample is chosen is known as the *population*, or *universe*. The results of analyses based on a sample generally extend beyond the sample itself by attempting to establish information concerning the population from which the sample is selected.

## 9-2    Statistic and Parameter

A measure computed from a sample is known as a *statistic* (not to be confused with "statistics," the body of techniques which we are studying). The mean age computed from a sample of students and other measures derived from samples, such as standard deviation and median age, are each examples of a statistic. Statistics are often used to estimate the corresponding *parameters*, which are measures of the population. We usually are interested in estimating a parameter on the basis of a statistic computed from a sample of the population.

Almost never will the statistic and the parameter be exactly identical. If a sample of 200 voters out of a total of 50,000 are asked to indicate their political preference, it would be most unlikely if the results based on the sample and on the universe were exactly the same. For example, the proportion expressing preference for the Republican party in a series of samples might be close to the population parameter, but seldom, if ever, identical. We would expect some difference, since each sample is affected by the particular group of persons being interviewed, and chance plays a role in the selection of these persons.

## 9-3    Reasons for Sampling

If the sample statistic is seldom a perfect estimate of the population parameter, one legitimately might ask, why use samples? There are several good reasons why sampling is used in statistical analysis. First, sampling saves time. It may be essential to obtain certain results within a limited amount of time, and an approximate result from sampling may be the only solution. Second, sampling may save money. Although obtaining an adequate sample often costs a considerable amount of money, it usually is much less than the cost of a complete census. Third, sampling may provide opportunities for control. Small samples may be given experimental treatments when such procedures could never be followed for the entire population. Fourth, sampling is sometimes more accurate than complete counts. For example, in the regular decennial census, the United States Bureau of the Census employs over 100,000 interviewers, each of whom has been given a short, intensive period of training. Under such conditions a large number of errors occur. It has been found that under certain circumstances more accurate results may be obtained by experienced and better-trained interviewers conducting a census on a sample basis. Finally, sampling is often the only way to procure relevant information. For example, in order to determine the efficiency of some new psychotherapeutic technique, it might be inexpedient to try it out on all the inmates in an institution, since there might be criticism for using an untried method on so large a group. A preliminary study based on a sample would provide adequate information to determine if a larger and more comprehensive study were warranted. In this connection it should be pointed out that a single institution is in turn a sample of all mental institutions.

Some people claim that they do not believe in sampling, that there is no way that one can be sure that the results of a sample provide adequate information. Without investigating the merits of their claim, it can be categorically stated that these people do believe in sampling methods. They give samples of their blood to a physician, rather than providing him with all their blood and then requesting a transfusion. They taste part of a cake icing, rather than eating it all before expressing an opinion. They thumb through a book before deciding whether to read it. They test the temperature of the water with their toes before plunging into the ocean. All this is done on the assumption that the sample will provide approximations to the population parameters.

## 9-4   Summary

A measure computed from a sample is a statistic.

A measure computed from a population or universe is a parameter.

Statistics are used to provide estimates of parameters.

# 10 Sampling Distributions

The drawing of successive samples of the same size produces a frequency distribution of measures computed from the samples. This frequency distribution is the basis of statistical inference.

## 10-1  Sampling Distributions

Since chance plays a role in determining what particular cases are included in each sample from a population, there inevitably will be some variability among sample statistics. If we took a series of samples of 30 students each from a school with a student population of 9,000 with a mean score in self-confidence of 50, the mean score in self-confidence for each sample would probably not be exactly 50. The statistic from the first sample might be 48, from the second 52, and from the third 51.

If this process were repeated dozens of times by taking successive samples of 30 students, some values of the statistics would occur more frequently than others. Values of the statistic close to the parameter of 50 would arise more often than statistics farther from 50. There would be a clustering of sample means around 50, with frequencies getting lower the farther the sample values diverge from 50. Sample means of 36 will occur much less frequently than sample means of 47, while 52 will be a sample result more often than 59.

A frequency distribution of sample means could be made from the results of the large number of successive samples of 30. The mean, standard deviation, and other measures could be computed for this frequency distribution. If the number of samples were infinitely large, the standard deviation would indicate the dispersion of sample means. The mean of the sample means would be the population mean, in this case 50. Frequency distributions of this kind are known as *sampling distributions*. They are frequency distributions of statistics from a large number of samples of a specified size. The standard deviation of these statistics is known as their *standard error*. (For an illustration of a sampling distribution, see Fig. 14-1.)

In later discussions, sampling distributions will be derived for various kinds of statistics, such as means, proportions, and correlation coefficients. Knowledge of such sampling distributions will answer many of the criticisms of sampling. Although sample results vary, the use of the standard error will aid in making generalizations about population parameters. For example, by knowing the standard deviation of sample means (the standard error of means), we have information as to the likelihood of certain amounts of deviation. With this knowledge, we can make generalizations with considerable precision. For example, if the sample mean is 48, under certain conditions we might be able to say with almost complete certainty that the population mean is between 45 and 51.

## 10-2   Efficiency

Some statistical measures are inefficient, unstable. *Efficiency* refers to the size of the standard error of a statistical measure. For example, the standard deviation is more efficient than the mean deviation because the standard error of the standard deviation is less than the standard error of the mean deviation. A small standard error means that sample statistics cluster closer to the population parameter. Therefore, a sample standard deviation is likely to be closer to the population standard deviation than the sample mean deviation is to its corresponding parameter. If we know the standard deviation and mean deviation of some sample, our estimate of the population parameter is likely to be closer for the standard deviation. The mean is the most efficient measure of central value, and the standard deviation is the most efficient measure of dispersion.

## 10-3    Random Sampling

In this elementary text, we always shall discuss samples that arise from one particular type of sampling method. If different processes of sampling are used, they will produce different sampling distributions. Accordingly, to simplify our discussion, all other processes of sampling will be ignored. All our computations of sampling distributions will be based on *random samples*, samples in which every case in the population has an equal chance of inclusion in the sample.

Random samples may not always give accurate estimates. A particular random sample might give a result that is far from the parameter being estimated. But our knowledge of the sampling distribution of random samples from a population will tell us how often such an unusually poor estimate will occur. The random process, in which there is equal likelihood of inclusion in the sample of each element in the population, provides a basis for deriving the sampling distribution.

Our ability to compute sampling distributions depends on the avoidance of human judgment in the choice of the sample. For example, suppose a sample of students were selected in order to estimate the median intelligence of the entire group. A person entrusted with the task of choosing the sample who was not acquainted with proper statistical procedures might do any of several different things. He might not include enough cases near the median, choosing extreme individuals of high and low intelligence in the belief that they would balance each other. He might leave out all extreme cases, choosing only average persons. Or he might think the persons of low intelligence were exceptional and not include any of them in the sample. No mathematical technique can take into account these twists and turns of human judgment. The sampling distributions can provide only for the operation of blind chance.

## 10-4   Problems in Random Sampling

It is usually very difficult to choose a random sample. For example, one possible method of choosing dwelling units in a city for inclusion in a sample of housing might involve the assignment of a number for each dwelling unit. Then numbers might be chosen from a rotating drum or from an already existing table of numbers which had been drawn at random. The dwelling units corresponding to the numbers selected would be included in the sample. This process appears random, but it is probably not perfectly so. It is unlikely that a list of all existing dwelling units can be made without considerable cost. New construction would probably be left out, meaning that new houses do not have an equal chance of being included in the sample.

Another illustration by which an apparently random method may give biased results involves the use of case materials from a social agency. It might seem that a random sample could be obtained by including each case whose folder appears at 1-inch intervals from the front of the file drawer. Since the more active cases have larger folders, there would be a greater likelihood of including them in the sample, thus resulting in a definite bias. On the other hand, some of the more active cases might inadvertently be excluded from the sample, since they would not be in the files, but on the desks of caseworkers.

In practice, the researcher does his best to select a random sample and points out the ways in which his sampling procedure diverges from true randomness. Minor deficiencies are unlikely to influence the results to any great extent, but experience is the main guide as to which departures from randomness might be considered minor.

## 10-5   Judgment Samples

Many samples, particularly those used by market-research agencies, are *judgment samples*, frequently referred to as "quota samples." The interviewer is told to interview a certain quota of persons of given characteristics, such as 40 low-income males between 30 and 40 years of age. Nonrandom elements immediately enter, since the choice of particular persons is left up to the judgment of the interviewer. It is possible that he might include more white-collar, better-educated persons in his sample, while correspondingly he might interview relatively few drunks, manual laborers, or sloppily dressed individuals. If his interviews were confined to residential areas during the day, he probably would have too many unemployed and too few workingmen. If he conducted his interviewing in the evening, it is likely that people who frequent places of amusement would tend to be excluded.

With rare exceptions, it is impossible to derive sampling distributions for judgment samples because of these factors of human judgment. The basic principles to be discussed in the analysis of random samples apply to other types of samples which use mechanical means of choosing a sample, but they cannot be applied to judgment samples.

## 10-6  Summary

Sampling distributions are frequency distributions of statistics.

The standard error is the standard deviation of a sampling distribution.

Efficiency refers to the size of the standard error of some statistic.

In random samples every case in the population has an equal chance of inclusion in the sample.

This text will discuss only sampling distributions arising from random samples.

It is almost never possible to derive the sampling distribution of a judgment sample.

# 11   Statistical Inference

Only three types of generalizations are possible from the results of a sample or group of samples to universe parameters. This chapter explains the logic of methods of statistical inference, reserving until later chapters the application of this logic to particular kinds of measures.

## 11-1   Statistical Inference and Decision Making

Statistical inference is concerned with making decisions. Unlike the hit-or-miss processes of decision making in social life, the rules for making decisions in statistical inference are specified in advance. If two statisticians agree on a common set of rules, they will make the same decisions. Although the rules are so designed that we can never be certain about the correctness of a particular decision, the proportion of incorrect decisions in the long run is known in advance. This puzzling statement will be clarified soon. For any particular decision a statistician never knows if he is right or wrong, but if a large number of decisions are made, he has a good idea how often he is wrong.

Imagine that a random sample of 10 adults is interviewed in a mining community. Every one of the 10 persons is male. What does this statistic tell us about the population parameter? Can we be sure that every person in the town is male? Obviously not, since this sample could be the result of chance. It is even theoretically possible, although extremely unlikely, that the 10 persons comprising the sample are the only males in town, and that every other person not in the sample is a female. We cannot make any decision about the parameter with absolute confidence (except, of course, that the town is not all female), since chance factors might occasionally produce samples which differ greatly from the population from which they are drawn.

It has now been shown that any decision we make about the proportion of males in the population may be wrong. For example, if we decide that the proportion of males in the population must be over .5, then we are saying that the sample did not come from a population with a male proportion of .5 or less. Although the population of a community may be half male and half female, it is possible to select a sample composed entirely of males because of the operation of chance factors.

If we wish to make decisions, we must be willing to make an occasional incorrect decision. For example, using the sampling distribution, suppose that we would obtain twice in 1,024 times a sample as unusual as this one from a population that is half male. In other words, 2 times out of 1,024 chance would produce an all-male or all-female sample of 10 cases when the parameter is actually .5. The odds are 1,022 (1,024 − 2 = 1,022) to 2 that this sample did not occur by chance from a population with equal numbers of each sex. If we reject the hypothesis that the true proportion of males is .5, we may be wrong but are probably right.

## 11-2   Type I Error

A *Type I error* refers to the rejection of a hypothesis when it should not be rejected. In the example above, there is a probability of .002 that a sample this unusual was drawn by chance from a population with a male proportion of .5. We rejected the hypothesis that the sample occurred by chance from such a population, and the probability of our being in error by rejecting this hypothesis is .002. In statistical inference we usually state in advance the proportion of Type I errors that we are willing to make.

## 11-3   Type II Error

If we never reject a hypothesis, we never shall make a Type I error. But such a practice will result in *Type II errors*, accepting a hypothesis when it should be rejected. Obviously we need a middle ground. In most social research, we are willing to make a Type I error 5 per cent of the time. The probability of a Type I error is therefore .05. Thus 1 time in every 20, or 5 times in every 100, the rejection of the hypothesis will be a mistake. The sample results will be due to chance, while we shall be denying that chance factors produced the observed statistics. In any particular case, there is no way of knowing whether the rejection of the hypothesis is correct or incorrect. But in the long run, there will be 19 correct rejections for every 1 that is incorrect.

| Decision | Actual situation | |
| --- | --- | --- |
| | Hypothesis true | Hypothesis false |
| Hypothesis true | Correct decision | Type II error |
| Hypothesis false | Type I error | Correct decision |

## 11-4   Level of Significance

The probability with which a Type I error is risked is known as *level of significance*. The level of significance is .05 in Sec. 11-3. Other levels of significance are used in certain situations, particularly .01 and .001. The choice of a significance level depends on the circumstances surrounding the decision to be made. For example, before rejecting the hypothesis that a bomb is about to explode, one would probably use a level of significance of .001. Even a 1 in 1,000 risk might not seem cautious enough.

## 11-5    Null Hypothesis

A hypothesis with a possibility of rejection at some level of significance is called a *statistical hypothesis*. The most widely used type of statistical hypothesis is the *null hypothesis*. In the type of example discussed above the null hypothesis describes the population from which a sample may have been drawn. The analysis proceeds to determine whether the discrepancy between the observed value and the population parameter according to the null hypothesis is so large that it is unlikely to be the result of chance. A second type of null hypothesis states that two samples come from the same population. The analysis then attempts to determine whether the observed difference in sample results could be due to chance.

If the probability of the null hypothesis being correct is less than .05, the null hypothesis is rejected. It will be seen that the Type I error is .05, since 1 out of 20 times the null hypothesis will have produced the sample statistic by chance when the null hypothesis should be rejected. The situation is more complex when the Type II error is analyzed. Failure to reject the null hypothesis does not mean that the null hypothesis is true. It merely means that the null hypothesis is one of many hypotheses which might produce the sample statistic by chance. Accordingly, if there is a probability greater than .05 that a null hypothesis is correct, we say that we fail to reject the null hypothesis. We do not accept the null hypothesis; we merely fail to reject it. In this text, the question of the probability of accepting a false hypothesis will not be discussed, although Type II errors are covered in more advanced statistical methods.

## 11-6    Testing Hypotheses about Relation of a Statistic to a Parameter

In the example of Sec. 11-1 the null hypothesis is that $p$ (the proportion) $= .5$. This hypothesis is rejected at the .05 level and the .01 level, but not at the .001 level. Since this text will almost always use the .05 level of significance, the null hypothesis is rejected.

If the proportion of males in the population were under .5, the deviation of the statistic from the hypothesized parameter would be even greater. The likelihood that this would occur by chance would be even less than .002. Therefore, the rejection of the null hypothesis that $p = .5$ automatically means that we reject all hypotheses that the parameter is less than .5, since these would be even more unlikely. Therefore, the small sample of 10 cases has clearly indicated that the proportion of males is above .5 in the mining town.

## 11-7   Testing Hypotheses about the Difference between Statistics from Two Samples

The second type of problem is very similar. It arises when we have two samples and there is a difference between the corresponding sample statistics. For example, the mean aggressiveness score of a class with a stern teacher may be 81, while the mean on aggressiveness for a class in the same grade with a sympathetic teacher may be 78. Is this difference *significant* (not due to chance), or is it the result of chance fluctuation?

If both samples came from the same population, they would not necessarily have exactly the same value for a statistic. If pairs of samples were repetitively taken from the same population, sometimes the first sample would be higher, and sometimes the second would have a higher value. The difference found by subtracting the second sample statistic from the first sample statistic would be negative half the time and positive half the time. The mean of these differences would be 0, since the sum of the differences is 0.

From the sampling distribution of the difference between two sample means, we compute the probability of a difference as unusual as 3. If a difference as large as 3 would occur  less than .05 of the time, then we reject the null hypothesis that the two samples come from the same population. There is a significant difference between the two samples. This does not necessarily mean that the difference is due to the difference in teachers, since there are other factors which might be producing differences. But we do believe that the two classes are markedly different with respect to aggressiveness.

If, on the other hand, the difference between sample means is not significant, then we fail to reject the hypothesis. Chance factors might produce the observed difference between samples.

## 11-8   Confidence Limits

The third and final type of problem in statistical inference is an application of the first type of problem to a series of null hypotheses. We have a sample statistic, such as the mean income of a group. We then wish to find the *95 per cent confidence limits,* an interval that probably contains the mean income (parameter) for the population from which this group is a sample. If the mean income in a sample is $4,000, the 95 per cent confidence limits might be $3,600 to $4,400. In other words, on the basis of a sample, the parameter is located within some interval.

The statement that these are 95 per cent confidence limits does not mean 95 per cent confidence that the population mean is between $3,600 and $4,400. Rather, it again refers to a repetitive process. For any particular set of confidence limits, the parameter may or may not be within the interval. But if the process of setting up such limits were repeated hundreds of times, the confidence interval would contain the parameter 95 per cent of the time.

The confidence limits arise from the rejection of whole sets of null hypotheses. The two parameters are found, one each side of the sample statistic, which would produce this particular sample value by chance exactly 5 per cent of the time. Every null hypothesis referring to a parameter more distant from the sample statistic than those two parameters would automatically be rejected, since the probability of chance occurrence of the statistic would be below .05. The sample value of $4,000 is significantly different at the .05 level from every parameter below $3,600 or above $4,400. Those two points form the 95 per cent confidence limits within which the parameter is believed to lie.

The 99 per cent confidence limits would be even wider, since we would want to be wrong only 1 time in 100 in saying that the parameter is between the limits. Unfortunately, the less willing we are to be wrong, the less precise must be our estimate of the interval containing the parameter.

## 11-9   Probability, Not Certainty

Obviously, the logic of statistical inference will only begin to be understood when applied to concrete situations. It should, however, be clear at this point that statistical inference never produces certainty. We do not know whether we are right or wrong in a decision to reject or not to reject a hypothesis, but we do know how often we shall be wrong in rejecting when we make a series of such decisions.

Some people may object that they want certainty, not probability. Such a demand is reasonable, but impossible of realization. Even in the physical sciences as a consequence of an increasing awareness of the limitations of measurement, the emphasis on the relativity and probability of scientific conclusions has supplanted the certainties of an earlier generation. In everyday activities, the reign of probability is also secure. When a traffic light turns green, cars and pedestrians start across the intersection because of the high probability that the light is red for the other stream of traffic. The teacher who calls on a student who raises his hand occasionally discovers that the student was merely in the process of scratching his head.

## 11-10   Use of Small Samples

The greatest development of modern statistics is the ability to use small samples with confidence. The rejection of a null hypothesis at the 5 per cent level of significance is just as definite with 25 cases as with 25,000. We have rejected the null hypothesis, and that is all that matters from a statistical point of view.

In fact, modern statistics often produces more realistic results with smaller samples. A very small difference causes a null hypothesis to be rejected when there are thousands of cases in the sample. In large samples chance factors play a very small role in the production of differences, so that most differences will be significant. But their statistical significance may have little relation to any social significance. For all practical purposes the differences may be unimportant even though not due to chance.

Finally, the use of a very large number of cases will not solve the problems of a biased sample. In the *Literary Digest's* Presidential poll of 1936, millions of ballots were mailed to people whose names appeared in telephone directories, automobile registries, and other listings. There was an underrepresentation of lower-income groups. As a result of this bias, as well as other factors, it was predicted that Landon would win the election. Actually Roosevelt won by an unprecedented landslide. In this connection it is interesting to note that the *Fortune Magazine* poll with a sample of 4,500 cases accurately predicted the popular vote for Roosevelt.

## 11-11    Summary

Type I error refers to the rejection of a hypothesis when it should not be rejected.

Type II error refers to accepting a hypothesis when it should be rejected.

Level of significance refers to the risk of a Type I error.

A null hypothesis may be rejected at some specified level of significance.

Three types of problems of statistical inference:

1. Does a sample come from a specified population?
2. Do two samples come from the same population?
3. Given a statistic, between what limits does the parameter lie?

# 12  The Binomial Distribution

VOCABULARY

probability
success ($p$)

failure ($q$)
mutually exclusive
independent
binomial distribution

This chapter is concerned with a common relative frequency distribution, the binomial. The binomial is the sampling distribution of proportions.

## 12-1  Probability

The exact meaning of *probability* is in dispute, but we shall define it simply as relative frequency over many trials. It is the proportion of all outcomes in which a particular event occurs, considered over the long run. For example, the probability of turning up heads when tossing a standard coin is .5, since one-half of all results will be heads if the coin is tossed several million times. In a class that is 60 per cent female, the probability of selecting a female for an interview through a random sampling process is .60. Although the probability of picking a female is .60, the selection of a single respondent may produce either a male or female. It should be noted especially that the concept of probability refers to relative frequency based on a very large number of trials, not to any particular case. Probability is concerned with the long run, a succession of many events.

Probabilities range from 0 to 1.0, since the numerator is the number of times a specific event occurs and the denominator is

132

the total of all events. A probability of 0 means that the particular event never occurs in an infinite number of trials, while a probability of 1.0 means that it always occurs. Most of the time, probabilities are proportions somewhere between such extreme values.

## 12-2   Success and Failure

Sometimes a certain type of event is labeled a *success*, with all other outcomes called *failures*. Any kind of event can be called a success, since the decision as to what is success and what is failure is completely arbitrary. For example, dying might be labeled success, while living would be correspondingly called failure. According to this nomenclature, an increase in the death rate would raise the probability of success. The point to remember is that the terms success and failure, as used in statistics, do not have the same meaning as in ordinary speech.

The probability of success plus the probability of failure always equals 1. If the probability of success is .4, then the probability of failure must be .6. The probability of success is usually symbolized by $p$, and the probability of failure is usually symbolized by $q$. Therefore, $p + q = 1$.

## 12-3   Mutually Exclusive Events

Sometimes two or more kinds of events are *mutually exclusive;* that is, if one event occurs then the others cannot occur. If we flip a coin, we get heads or tails, not both. A person cannot be both a Democrat and a Republican. A candidate is elected or he is defeated. These pairs of events are mutually exclusive.

The probability of obtaining either of two or more mutually exclusive events is the sum of their probabilities. If no one can have two majors, and the proportion of English majors is .10 and of sociology majors is .03, then the probability of selecting either a sociology or an English major is .13. Since success and failure are mutually exclusive, the probability of obtaining either a success or a failure is equal to the sum of their probabilities, or 1.

## 12-4   Independent Events

Sometimes probabilities are *independent*, neither probability being affected by the outcome of the other event. If a coin turns up heads, it does not affect the next toss. With independent probabilities, the probability of two events occurring is equal to the product of their probabilities. If the probability of guessing the right answer on a test is .50, the probability of guessing correctly on three successive questions is

$$(.50)(.50)(.50) = .125$$

## 12-5   Probability of Obtaining a Sample Result

If we select a random sample from some population in which the proportion of females is .60, what is the probability of obtaining a sample with six females and one male? To solve this problem, let us imagine that six females were chosen successively, followed by a single male. This sample could be represented as follows:

$$F \quad F \quad F \quad F \quad F \quad F \quad M$$

Each of these choices is independent, so that the probability of their occurring simultaneously in the same sample is equal to the product of their probabilities.

$$\begin{aligned}
\text{Probability of this sample} &= (.6)(.6)(.6)(.6)(.6)(.6)(.4) \\
&= (.6)^6(.4) \\
&= .0187
\end{aligned}$$

We now have determined the probability of obtaining this particular sample, six females in a row and then one male, from a population in which the proportion of females is .60. There are seven different ways by which six females and one male might be included in a random sample, but we have ascertained the probability of selecting only one of these seven different samples. The following portrays the pattern of seven samples in which each is composed of one man and six women:

| | | | | | | | |
|---|---|---|---|---|---|---|---|
| 1. | F | F | F | F | F | F | M |
| 2. | F | F | F | F | F | M | F |
| 3. | F | F | F | F | M | F | F |
| 4. | F | F | F. | M | F | F | F |
| 5. | F | F | M | F | F | F | F |
| 6. | F | M | F | F | F | F | F |
| 7. | M | F | F | F | F | F | F |

Let us find the probability of the fourth sample.

$$\text{Probability of fourth sample} = (.6)(.6)(.6)(.4)(.6)(.6)(.6)$$
$$= (.6)^6(.4)$$
$$= .0187$$

It can be seen that each of the samples has the same probability, .0187. They are mutually exclusive, since, if one sample were drawn, another one cannot be drawn at that time. Therefore, the probability of obtaining any one of these seven samples is the sum of their separate probabilities, or $7(.0187) = .1309$. This is the answer to our original question.

## 12-6  Binomial Distribution

There is a more direct method of solving problems of this type. The *binomial distribution*, which is $(p + q)^N$, will quickly provide the correct answer. If $p$ is the proportion of females and $q$ the proportion of males in the population and $N$ is the number of cases in the sample, then we have $(p + q)^7$ as the distribution of probabilities for this problem. This is equal to

$$(p + q)(p + q)(p + q)(p + q)(p + q)(p + q)(p + q)$$

The result of these multiplications will be

$$p^7 + 7p^6q + 21p^5q^2 + 35p^4q^3 + 35p^3q^4 + 21p^2q^5 + 7pq^6 + q^7$$

$$(.6)^7 + 7(.6)^6(.4) + 21(.6)^5(.4)^2 + 35(.6)^4(.4)^3 + 35(.6)^3(.4)^4$$
$$+ 21(.6)^2(.4)^5 + 7(.6)(.4)^6 + (.4)^7$$

The exponents indicate the number of males and number of females to which the particular probability refers. The second term of this result has $p$ to the sixth power (six females) and $q$

to the first power (one male). Therefore, the value of this term will tell us the probability of obtaining six females and one male in a sample.

$$7p^6q = 7(.6)^6(.4)$$
$$= 7(.0187)$$
$$= .1309$$

The sum of all the terms after we have expanded $(p + q)^N$ must be equal to 1.0, since the sum of all possible probabilities must be unity. This requirement is met by the binomial distribution, since $p + q$ must equal 1, and 1 to any power $(N)$ is always 1.

The only difficulty in using the binomial distribution to find probabilities is the necessity of writing all the terms in the expansion. Close examination of the above will disclose regularities that enable us to write the binomial expansion without doing any multiplying. Observe the first term $(p^7)$. The power of $p$ in the first term is $N(N = 7)$, while the power of $q$ is $0(q^0 = 1)$. In each succeeding term, the power of $p$ diminishes one unit. At the same time, the power of $q$ increases by 1 in each successive term. The pattern of $p$ and $q$ is clear, with the power of $p$ decreasing by one unit each time and the power of $q$ increasing by one unit. The sum of the powers of $p$ and $q$ in each term must be $N$. The next problem is to find a regularity in the coefficients (the number of possible combinations that will have a specified number of males and females). The coefficient in front of the first term is 1, which we do not bother to write. Multiplying the coefficient by the power of $p$ and dividing that result by 1 more than the power of $q$ will give the coefficient of the next term. The coefficient (1) times the power of $p$ (7) gives 7. The power of $q$ is 0, so that 1 more than the power of $q$ is 1. Dividing 7 by 1 gives 7, the coefficient of the second term.

**Illustration.** Expand $(p + q)^5$

First write the powers of $p$, descending one unit for each term

$$p^5 \quad p^4 \quad p^3 \quad p^2 \quad p$$

Include the power of $q$ in each term, adding one unit and starting with $q^0$.

$$p^5 \quad p^4q \quad p^3q^2 \quad p^2q^3 \quad pq^4 \quad q^5$$

We next insert the coefficients, starting with 1.   One multiplied by 5, the power of $p$, is divided by 1 more than the power of $q$. The result is 5, the next coefficient.

The binomial expansion now is

$$p^5 + 5p^4q$$

The coefficient of the next term will be 5(4), divided by (1 + 1).

$$p^5 + 5p^4q + 10p^3q^2$$

The next coefficient will be 10(3) divided by (2 + 1), or 10.

$$p^5 + 5p^4q + 10p^3q^2 + 10p^2q^3$$

Can you find the next two terms?

We can use the binomial distribution as our sampling distribution in order to test a null hypothesis about the proportion of some event in the universe. The binomial expansion will aid in determining how often a sample as unusual as the one observed would come from some hypothesized population. Figure 12-1 shows how closely the binomial distribution approximates an actual sampling distribution.

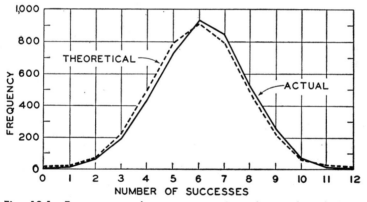

Fig. 12-1  Frequency polygons portraying observed and theoretical distributions based on the throwing of twelve dice 4,096 times. (*Original data from G. Udny Yule and M. G. Kendall, An Introduction to the Theory of Statistics, 1937, p. 424.*)

**Illustration.** In a sample of six persons all say that they oppose a certain bill in Congress. On the basis of this small random sample, is it proper to conclude that most of the people comprising this particular population are opposed to the bill?

The binomial expansion is

$$p^6 + 6p^5q + 15p^4q^2 + 20p^3q^3 + 15p^2q^4 + 6pq^5 + q^6$$

Before we observed our sample result, we did not state that we expected most of the population to oppose this bill. We would not have questioned a sample result in the opposite direction, all supporting the bill. Therefore, $p^6 + q^6$ will give the total probability of obtaining sample results this unusual when the proportion opposing the bill is .5.

$$\begin{aligned} p^6 + q^6 &= (.5)^6 + (.5)^6 \\ &= .0156 + .0156 \\ &= .0312 \end{aligned}$$

*Decision:* Since a result as unusual as this would occur by chance less than 5 per cent of the time, we reject the null hypothesis. If the null hypothesis were to indicate that the population proportion opposing the bill was some particular number below .5, the likelihood of this result by chance would be even lower, and the hypothesis would be rejected. Having rejected all such hypotheses, we conclude that most of the people in this community oppose this bill.

## 12-7   Summary

Probability is relative frequency over many trials.

The binomial distribution gives the sampling distribution of proportions.

## PROBLEMS FOR CHAP. 12

1. Alfred Strongheart considers himself very intelligent, and he challenges Alibi Ike to a contest. They each attempt to do *The New York Times* crossword puzzle on six successive Sundays. Alfred makes a better showing four times out of six. Alibi Ike defends his honor by saying that Alfred is just winning by luck. Is this a justifiable position?

*Ans.* Ike has a good alibi

2. If the probability of recidivism for persons now in the state penitentiary is .60, what is the likelihood of their "going straight"?

*Ans.* .40

3. In eight discussion groups composed of equal proportions of males and females, it is observed that the spontaneous leader is a male in every case. What proportion of the time would a result this unusual occur by chance?

*Ans.* .008

# 13 The Normal Distribution

VOCABULARY        normal curve
                  kurtosis
                  mesokurtic
                  platykurtic
                  leptokurtic

This chapter leads from the binomial distribution to the normal curve. The normal curve is emphasized because the sampling distributions of several statistics take that characteristic form.

## 13-1 Binomial Distribution and the Normal Curve

Figures 13-1 and 13-2 show the probability of obtaining each specified number of successes in repeated samples of 4, 6, 8, and 10 cases each where the population proportion is .5. Each number of successes is thought of as the mid-point of an interval from .5 below the number to .5 above the number. The number 7, for example, is the mid-point of the class 6.5 to 7.5. Discrete results are treated as if they were continuous, making it possible to construct histograms and curves based on relative frequencies.

As the number of cases in the sample increases, from 4 to 6 to 8 to 10, there is a gradual smoothing of the curve. The frequency distribution approaches the characteristic bell shape of the *normal curve*. This same tendency for the binomial distribution to approach the normal curve is found whenever the number of cases in each sample increases, regardless of the value

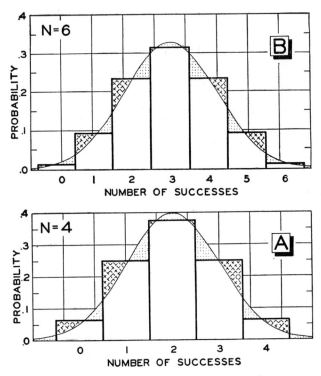

Fig. 13-1    Probability for specified number of successes
in different sized samples. (A) sample of four cases; (B)
sample of six cases.

of $p$ and $q$. Some of the other sampling distributions which we
shall study are normal in shape, and consequently there will be
considerable application of the normal curve in statistical inference.

The formula for the normal curve, which you do not have to
learn, is

$$Y = \frac{1}{\sigma \sqrt{2\pi}} \, e^{-x^2/2\sigma^2}$$

It is worthy of note that the only variables in the equation are
$Y$ and $x$. $\sigma$ is constant for any given sample, while $e$ and $\pi$ are
always 2.7183 and 3.1416, respectively. Therefore, the only

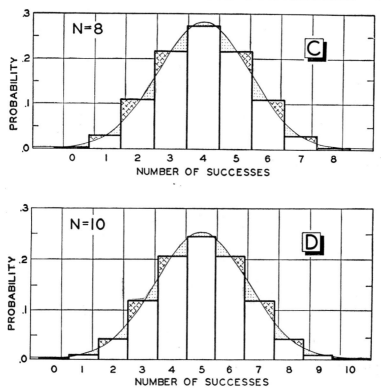

Fig. 13-2   Probability for specified number of successes in different sized samples. (*C*) sample of eight cases; (*D*) sample of ten cases. It will be observed from Figs. 13-1 and 13-2 that there is a gradual smoothing of the frequency curve as the size of the sample is increased.

variable affecting the relative frequency is the deviation from the mean.

## 13-2   Normal Curve and Standard Score

The normal curve would have much wider utility if the deviations from the mean were in terms of some pure number, independent of the units in any particular analysis. This is accomplished by using standard scores, $z = x/\sigma$. The result is our ability to use the normal curve in a wide variety of situations, with standard scores represented by the $X$ axis.

Since the normal curve is based on standard scores, we always know its mean and standard deviation. As in any group of standard scores, the mean is 0 and $\sigma = 1$.

**Proof.** That the mean of a group of standard scores is 0.

$$\bar{X} = \frac{\Sigma \frac{x}{\sigma}}{N} \qquad \text{by definition of the mean as } \Sigma X/N$$

$\sigma$ is a constant so that

$$\bar{X} = \frac{(1/\sigma)\,\Sigma x}{N} \qquad \text{by Rule 2}$$

But $\Sigma x$ always must equal 0. Therefore

$$\bar{X} = \frac{0}{N} = 0$$

**Proof.** That the standard deviation of a group of standard scores must be equal to 1.

$$\sigma = \sqrt{\frac{\Sigma(X - \bar{X})^2}{N}}$$

For standard scores, in which the mean is 0,

$$\sigma = \sqrt{\frac{\Sigma\left(\frac{x}{\sigma} - 0\right)^2}{N}}$$

$$= \sqrt{\frac{\Sigma \frac{x^2}{\sigma^2}}{N}}$$

$$= \sqrt{\left(\frac{1}{\sigma^2}\right)\frac{\Sigma x^2}{N}}$$

We can take the square root of $1/\sigma^2$. Its square root is $1/\sigma$. This can be placed in front of the radical sign.

$$\sigma = \frac{1}{\sigma}\sqrt{\frac{\Sigma x^2}{N}}$$

By definition,

$$\sigma = \sqrt{\frac{\Sigma x^2}{N}}$$

Substituting into the previous equation,

$$\sigma = \frac{1}{\sigma}\,(\sigma)$$
$$= 1$$

## 13-3    Area under the Normal Curve

The total area under the normal curve, the sum of all the probabilities, is equal to 1. Table B in the Appendix gives the area between the mean and the standard score. The proportion of frequencies below the mean is equal to .5, as is the proportion above the mean. This is true because the normal curve is perfectly symmetrical, as can be seen in Figs. 13-3 and 13-4.

Since the normal curve will be used in much of the work in statistical inference, some of its most important characteristics are worth examination. For example, in a normal distribution a little over 2/3 of the cases will lie less than one standard deviation from the mean. By referring to Table B it will be noted that .3413 of the cases are between the mean and a point one standard deviation from the mean. Since the normal curve is symmetrical, .3413 of the cases also lie on the other side of the mean up to a point one standard deviation away. Therefore, in a normal distribution 68.26 per cent of the cases are between $z = -1$ and $z = 1$, deviations of less than one standard deviation from the mean (Fig. 13-3).

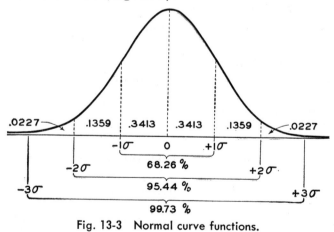

Fig. 13-3    Normal curve functions.

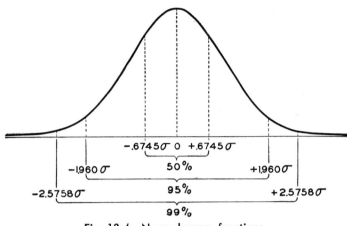

Fig. 13-4  Normal curve functions.

There is one number in Table B that is worth memorizing. It is 1.96. In a normal distribution a deviation greater than $1.96\sigma$ would occur less than 5 per cent of the time, so that we use this value as the boundary beyond which we reject the null hypothesis (Fig. 13-4). In the body of the table, .4750 is the reading for $x/\sigma = 1.96$. This indicates that 47.5 per cent of the cases are between that point and the mean. It also implies that 2.5 per cent of the cases are beyond this point (50 per cent are above the mean, and $50 - 47.5 = 2.5$). Using the two tails of the normal curve, values as unusual as 1.96 would occur exactly 5 per cent of the time (2 times 2.5). If a sampling distribution forms a normal curve, a deviation of a sample value from a population parameter that is greater than $1.96\sigma$ is going to occur less than 5 per cent of the time by chance. Therefore, this difference is significant at the .05 level, and the null hypothesis is rejected.

**Problem.** In a normal curve,

a. What proportion of the cases fall between the mean and $z = -.2$?
*Ans.* .0793

b. What proportion of the cases fall between the mean and $z = 1.75$?
*Ans.* .4599

c. What proportion of the cases deviate from the mean as much as or more than $z = 1.8$?   *Ans.* .0718

d. What proportion of the cases are above $z = 1.4$?   *Ans.* .0808 *e.* What proportion of the cases are between $z = -1.0$ and $z = .6$?

*Ans.* .5671

*f.* What proportion of the cases are between $z = 1.6$ and $z = 1.92$?

*Ans.* .0274

*Hint:* Make a rough sketch in each problem of the area you are interested in finding.

## 13-4 Skewness and the Normal Curve

In the normal curve, the mean, median, and mode coincide. This fact follows because the normal curve is perfectly symmetrical. When distributions are not symmetrical, we say they are skewed. Skewness may be either positive or negative. Since the mean is greatly influenced by cases at the extremes, it will be found farther in the direction of skew than either the median or the mode. Figure 13-5 illustrates the characteristic positions of the mean and median in a positively skewed distribution.

## 13-5 Kurtosis and the Normal Curve

Some curves are more peaked than others, have a greater concentration of cases near the mode. *Kurtosis* is the term applied to the degree of peakedness. The normal curve is used as the standard, with its peakedness labeled *mesokurtic.* A distribution which is less concentrated, flatter than the normal

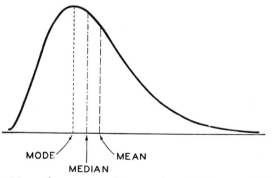

Fig. 13-5 Position of mean, median, and mode in a positively skewed distribution.

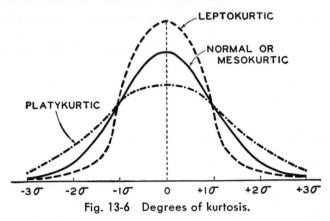

Fig. 13-6   Degrees of kurtosis.

curve, is called *platykurtic*. A curve more peaked than the normal curve is designated as *leptokurtic* (Fig. 13-6).

## 13-6   Summary

Large samples smooth the binomial and produce the normal curve.

The normal curve is measured in standard scores on the $X$ axis, with the area under the curve representing probability.

For a normal curve, a deviation greater than 1.96 in standard score would occur less than 5 per cent of the time.

Kurtosis refers to relative degree of peakedness of a distribution.

### PROBLEMS FOR CHAP. 13

**1.** Several thousand entering freshmen at a large state university take a vocabulary test. The scores on this test are normally distributed with a mean of 65 and standard deviation of 8. Fitzhugh Abercrombie has a score of 77 on the vocabulary test. What proportion of the students did better than he?                    *Ans.* .0668

**2.** In a normal distribution, what proportion of the cases fall below $z = -1.96$?                    *Ans.* .025

**3.** In a normal distribution, what proportion of the cases fall between $z = -1.80$ and $z = -1.77$?                    *Ans.* .0025

# 14 Sampling Distribution of Proportions

Previous chapters have developed the general background of statistical inference. In this chapter the logic of testing hypotheses is illustrated by analysis of the sampling distribution of proportions.

## 14-1 Sampling Distribution of Proportions and the Normal Curve

In order to avoid confusion, let us designate as $p_u$ and $q_u$ the *universe proportions* which have or do not have, respectively, a certain characteristic. Correspondingly, $p_s$ and $q_s$ will be used to symbolize *sample proportions*. If we selected a large number of random samples, the mean of the sample proportions would be nearly equal to the population proportion.

The sampling distribution of proportions, the result of taking a large number of samples of equal size, tends to be approximately normal if $Np_u$ and $Nq_u$ are both over 5 (see Chap. 13). If, however, the product of the number of cases in the sample and either universe proportion is not greater than 5, the binomial distribution, not the normal, should usually be used for the sampling distribution (Chap. 12).

Since we are dealing with qualitative data with respect to proportions, this illustrates that a sampling distribution can be normal although the underlying distribution is not normal. This is frequently the case; the sampling distribution is normal in many situations in which the samples themselves are far from normal. This does not, however, mean that all sampling distributions are normal.

The sampling distribution of proportions is based on a great many samples. The mean of the sampling distribution, we have noted, is $p_u$. Since we know the mean of the sampling distribution and the proportion in a sample, we can determine the probability of deviations between $p_u$ and $p_s$ by converting the values into standard score units and using Table B in the Appendix.

## 14-2   Standard Error of Proportions

The *standard error of proportions*, which is the standard deviation of sample proportions around the population proportion, is symbolized by $\sigma_p$.

$$\sigma_p = \sqrt{\frac{p_u q_u}{N}}$$

With this additional information, we can now test hypotheses.

**Illustration.** In a recent election in a large city, it was found that the proportion who voted in favor of a school bond issue was 62 per cent. One week later a random sample of 100 persons was interviewed, and 71 per cent indicated that they were in favor of the school bond issue. What does this difference signify? Is this difference the result of sampling variability due to chance, or is it the result of band-wagon psychology, where people tend to identify themselves with the winning side?

$$N p_u = 100(.62) = 62 \quad \text{which is over 5}$$
$$N q_u = 100(.38) = 38 \quad \text{which is over 5}$$

Therefore, we can use the normal curve as the approximate sampling distribution.

*Null hypothesis:* This is a random sample from a population whose proportion is .62. The deviation of $p_s$ from $p_u$ is due to sampling variability. The standard error of proportions is

$$\sigma_p = \sqrt{\frac{p_u q_u}{N}}$$

$$= \sqrt{\frac{(.62)(.38)}{100}}$$

$$= \sqrt{.002356}$$

$$= .0485$$

Computation of $z$ score:

$$z = \frac{.71 - .62}{.0485}$$

$$= \frac{.09}{.0485}$$

$$= 1.85$$

Probability that a result this unusual would happen by chance:

From Table B, .4678 of the area is between 1.85 and the mean. The probability of obtaining a $z$ score above 1.85 is (.5000 − .4678), or .0322. The probability of a value so unusual (using both tails of the normal distribution) is 2(.0322), or .0644.

*Decision:* Since a sample this unusual could arise from sampling variability more than 5 per cent of the time from the stated population, we cannot reject the null hypothesis. No significant effect has been demonstrated.

## 14-3    Difference between Two Sample Proportions

The second type of problem to be investigated involves the significance of differences between results from two samples. If we develop the sampling distribution of differences between two sample proportions, the sampling distribution of such differences forms a normal curve. The mean of the sampling distribution is always zero since, in the long run, the first sample proportion would be higher than the second 50 per cent of the time, and the second sample proportion would be higher in 50 per cent of the comparisons.

In order to apply the normal curve of areas, the difference between two sample proportions must be converted into $z$ scores.

The mean of the sampling distribution is known to be 0, but the standard error of the difference between two proportions has not yet been discussed.

The standard error of the difference between any two independent measures is equal to the square root of the sum of the squares of their standard errors. For proportions from large samples,

$$\sigma_{p_1-p_2} = \sqrt{\sigma_{p_1}^2 + \sigma_{p_2}^2}$$

The formula above uses $p_1$ and $p_2$ for the universe proportions in the two populations from which the samples are drawn. These are, of course, unknown. A solution to this problem arises from our interest in determining whether there is a significant difference between the two sample proportions. In all such situations our null hypothesis is that there is no difference between the population proportions. Therefore, $p_u = p_1 = p_2$.

Substituting in the formula above,

$$\sigma_{p_1-p_2} = \sqrt{\frac{p_u q_u}{N_1} + \frac{p_u q_u}{N_2}}$$

But $p_u$ is unknown. We shall approximate it by the relative frequency of successes when both samples are combined. This will give a better estimate of $p_u$ than either one of the two samples. The estimated population proportion is symbolized by $\hat{p}_u$.

One might suppose that the use of $\hat{p}_u$ in place of $p_u$ would introduce considerable inaccuracy. In most cases, however, $\hat{p}_u \hat{q}_u$ usually gives results very close to the correct $p_u q_u$. For example, let $\hat{p}_u$ be .6, and the true population proportion be .7. Then $\hat{p}_u \hat{q}_u = (.6)(.4) = .24$, while $p_u q_u = (.7)(.3) = .21$. The difference is only .03 in this example in which $\hat{p}_u$ is .10 away from $p_u$.

**Illustration.** A group of 100 men are compared with a group of 200 women with respect to their attitude toward compulsory military training in high school. Among the men, 45 per cent favor such training, as compared to 30 per cent of the women who express a favorable attitude toward military training. Does this difference actually indicate that men are more favorably disposed to high-school military training than women?

*Null hypothesis:* The two samples come from two populations with exactly the same proportion favoring military training. The difference between the sample proportions is due to sampling variability, with the mean difference equal to 0. The standard error of the difference between proportions is

$$\sigma_{p_1-p_2} = \sqrt{\frac{\hat{p}_u\hat{q}_u}{N_1} + \frac{\hat{p}_u\hat{q}_u}{N_2}}$$

Among the men, 45 favor compulsory military training, since

$$(.45)(100) = 45$$

Among the women, 60 favor compulsory military training, since

$$(.30)(200) = 60$$

Combining the two samples to obtain an estimate of the universe proportion,

$$\hat{p}_u = \frac{45 + 60}{100 + 200}$$
$$= {}^{105}\!/_{300}$$
$$= .35$$
$$\hat{q}_u = 1 - .35$$
$$= .65$$
$$\sigma_{p_1-p_2} = \sqrt{\frac{(.35)(.65)}{100} + \frac{(.35)(.65)}{200}}$$
$$= \sqrt{\frac{.2275}{100} + \frac{.2275}{200}}$$
$$= \sqrt{.002275 + .001138}$$
$$= \sqrt{.003413}$$
$$= .058$$

Computation of $z$ score:

$$z = \frac{(.45 - .30) - 0}{.058}$$
$$= \frac{.15}{.058}$$
$$= 2.59$$

Probability that a result this unusual would happen by chance:

The sampling distribution of differences between proportions forms a normal curve. From Table B, .4952 of the area is between 2.59 and the mean. The probability of a value so unusual (using both tails of the normal curve) is 2(.0048) = .0096. Since $z$ was larger than 1.96, the probability had to be less than .05.

*Decision:* Since a difference between sample proportions this unusual would arise less than 5 per cent of the time, we reject the null hypothesis. Men tend to favor compulsory military training in high school to a greater extent than women.

## 14-4   Confidence Limits of Proportions

The final type of problem involves the estimation of the population proportion when the sample proportion is known. We shall find the 95 per cent confidence limits which are expected to contain $p_u$. This is done by finding the population proportions which would give the observed sample result by chance exactly 5 per cent of the time. Any population proportions more distant from the sample proportion would produce a sample this unusual, or more unusual, somewhat less than 5 per cent of the time, so we can reject such proportions.

In finding the confidence limits for the population proportion, it is necessary to use the standard error of proportions. The formula for the standard error of proportions requires the use of $p_u$, although we have only $p_s$, the proportion in the sample. It is necessary to substitute $p_s$ for $p_u$, since it is the best available estimate of the population proportion. Obviously, this introduces some error, but the discrepancy is usually slight. In general, $p_s q_s$ gives almost the same result as $p_u q_u$ (Sec. 14-3).

**Illustration:** In a random sample of 150 patients at a large mental hospital, 43 per cent had no brothers or sisters. What are the 95 per cent confidence limits for the proportion of only children at the hospital? The standard error of proportions is

$$\sigma_p = \sqrt{\frac{\hat{p}_u \hat{q}_u}{N}}$$

$$= \sqrt{\frac{(.43)(.57)}{150}}$$

$$= \sqrt{\frac{.2451}{150}}$$

$$= \sqrt{.001634}$$

$$= .0404$$

Computation of 95 per cent confidence limits:

The $z$ score for the deviation of a sample proportion from a population proportion must be exactly 1.96, when a $z$ score so unusual or more unusual would occur 5 per cent of the time. A standard score greater than 1.96 would occur less than 5 per cent of the time, and deviations that large would require us to reject the hypothesis that the sample came from this population.

$$1.96(.0404) = .079$$

The smallest population proportion likely to give us this sample proportion is $.43 - .079 = .35$. The largest proportion would be

$$.43 + .079 = .51$$

The 95 per cent confidence limits are .35 to .51.

The proportion of only children in the mental hospital is probably between .35 and .51.

If we were to make a series of such estimates, we would be wrong approximately 5 times out of 100.

## 14-5  Standard Error and Number of Cases

Since $N$ appears in the denominator under the square-root sign, it is apparent that an increase in the size of a sample will decrease the standard error. This is true for all types of measures. However, since the square root of $N$ is in the denominator, the standard error does not decrease in direct relation to the number of cases. Since the square root of 4 is 2, it takes 4 times as many cases to reduce the standard error by 1/2. (See Fig. 14-1 to observe the decrease in standard error with an increase in sample size.)

It is significant to note that in statistical inference our concern is with the size of the sample, not with the size of the population to which we generalize. As long as the population is more than 5 times the size of the sample, there need be no interest in the size of the universe. A sample of 50 cases can be used for statistical inference for a population of 500 cases just as well as for a population of 1,000,000.

When dealing with proportions, it is relatively easy to determine before the sample is drawn the maximum number of cases in order to obtain a certain degree of accuracy. This can be done

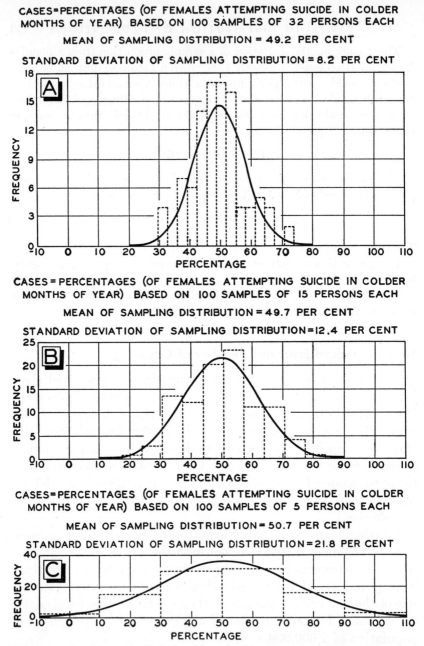

CASES=PERCENTAGES (OF FEMALES ATTEMPTING SUICIDE IN COLDER MONTHS OF YEAR) BASED ON 100 SAMPLES OF 32 PERSONS EACH

MEAN OF SAMPLING DISTRIBUTION = 49.2 PER CENT

STANDARD DEVIATION OF SAMPLING DISTRIBUTION = 8.2 PER CENT

CASES = PERCENTAGES (OF FEMALES ATTEMPTING SUICIDE IN COLDER MONTHS OF YEAR) BASED ON 100 SAMPLES OF 15 PERSONS EACH

MEAN OF SAMPLING DISTRIBUTION = 49.7 PER CENT

STANDARD DEVIATION OF SAMPLING DISTRIBUTION = 12.4 PER CENT

CASES=PERCENTAGES (OF FEMALES ATTEMPTING SUICIDE IN COLDER MONTHS OF YEAR) BASED ON 100 SAMPLES OF 5 PERSONS EACH

MEAN OF SAMPLING DISTRIBUTION = 50.7 PER CENT

STANDARD DEVIATION OF SAMPLING DISTRIBUTION = 21.8 PER CENT

Fig. 14-1 Sampling distributions of proportions based on 100 samples of 32, 15, and 5, respectively. These samples are derived from a universe of 643 female attempted suicides. The data indicate the proportion of attempted suicides occurring during the colder months (October to March) of the year. For the universe, 49.12 per cent of the cases were reported during this part of the year. It will be noted that the mean percentages for the three sampling distributions are very similar.

because the maximum standard error will occur if $p_u = .5$. If $p_u$ is not .5, either higher or lower, the standard error will be less. Therefore, we can compute the maximum number of cases, since it will be more than adequate for all other situations.

**Illustration.** We want to be 95 per cent confident that our estimate is within .1 of the true proportion. What is the maximum number of cases needed in the sample?

The sample must not be over .1 from the true proportion, so that .1 must be the distance from the sample result to the 95 per cent confidence limits.

$$.1 = 1.96\sigma_p$$

Therefore, dividing both sides by 1.96,

$$\sigma_p = \frac{.1}{1.96}$$
$$= .051$$

Using $p_u$ as .5 to obtain a maximum standard error,

$$.051 = \sqrt{\frac{(.5)(.5)}{N}}$$
$$= \sqrt{\frac{.25}{N}}$$

Squaring both sides,

$$(.051)^2 = \frac{.25}{N}$$
$$.002601 = \frac{.25}{N}$$

Multiplying both sides by $N$,

$$.002601N = .25$$

Dividing both sides by .002601,

$$N = 96.1$$

A sample of 97 always will allow us to be 95 per cent confident that our sample proportion is within .1 of the population proportion.

## 14-6  Summary

The sampling distributions of proportions are approximately normal if $Np_u$ and $Nq_u$ are both over 5.

$$\sigma_p = \sqrt{\frac{p_u q_u}{N}}$$

The sampling distribution of differences between proportions tends to be normal, with

$$\sigma_{p_1-p_2} = \sqrt{\frac{p_u q_u}{N_1} + \frac{p_u q_u}{N_2}}$$

Statistical inference is concerned only with the size of the sample, not with the size of the population, as long as the population is more than five times the size of the sample.

By setting $p_u$ at .5, it is possible to derive the maximum number of cases necessary for some specified accuracy at a certain level of confidence.

## PROBLEMS FOR CHAP. 14

**1.** In 1950 the proportion of males over 14 years of age in the labor force in the United States was 82.4 per cent. In a large Northern city, the proportion was only 80.5 per cent for the 200,000 males over 14. If we think of the data as coming from a random sample, is this small difference statistically significant?                                   *Ans.* Yes

**2.** A sample of 200 older women are asked whether they oppose marriages in which the female is older than the male. The same question is asked of 100 female college students. The proportion opposed among the college girls is .34, while only .18 of the older women object to such marriages. Is this difference due to random fluctuations in sample responses?                                   *Ans.* No

**3.** In a post–World War II survey, 39 per cent of a sample of 1,000 Germans express negative feelings toward "hypocrites." What are the 95 per cent confidence limits for the proportion in the population having this attitude?                                   *Ans.* .36 to .42

# 15 Sampling Distribution of Means

population mean ($\mu$)
sample mean ($\bar{X}$)
standard error of means ($\sigma_{\bar{x}}$)

universe standard deviation ($\sigma$)
sample standard deviation ($s$)
bias
degrees of freedom
Student's $t$ ($t$)
standard error of the difference between
two means ($\sigma_{\bar{X}_1 - \bar{X}_2}$)

In this chapter the logic of statistical inference, previously applied to proportions, is applied to sample means. Two related methods of analysis are discussed, the choice of method depending on the number of cases in the sample.

## 15-1 Sampling Distribution of Means and Normal Curve

If the number of cases in a series of samples is 30 or more, the sampling distribution of means will be approximately normal. Since most social characteristics are not distributed normally, it is of considerable advantage that samples with 30 or more cases actually form a normal sampling distribution of means although the underlying distribution in the universe is not normal. Income, for example, invariably manifests marked positive skewness, but the sampling distribution of mean incomes of random samples of 30 or more cases will tend to be normal.

## 15-2 Standard Error of Means

The mean of the sampling distribution of sample means will be approximately the same as the *population mean,* to be symbolized

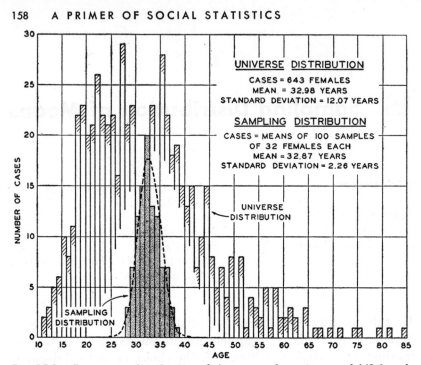

**Fig. 15-1.** Frequency distribution of the ages of a universe of 643 female attempted suicides and means of 100 random samples of 32 cases each. Note especially the form of the sampling distribution, as well as its relatively small standard deviation. It also should be observed that there is only a slight difference between the mean of the universe (32.98 years) and the mean of the sampling distribution (32.87 years).

by $\mu$ (Greek mu). The *standard error of means* ($\sigma_{\bar{x}}$) is found by dividing the standard deviation of the cases in the universe by the square root of the number of cases in the sample.

$$\sigma_{\bar{x}} = \frac{\sigma}{\sqrt{N}}$$

Note that the standard error refers to the deviation of sample means around the population mean, while the standard deviation refers to the deviation of individual values around the population mean. Figure 15-1 shows both the universe distribution and sampling distribution on the same chart.

The formula above requires that we know the *standard deviation of the universe* ($\sigma$). Since we are engaged in statistical inference from sample data, most of the time we do not know the standard deviation of the population. We must estimate the standard deviation of the universe from the *standard deviation of a sample*, which is symbolized as $s$.

## 15-3    Bias of Sample Standard Deviation

The standard deviation of a sample tends to be smaller than the standard deviation of the universe. The sum of the squared deviations around the mean is a minimum (Sec. 5-7), so that the standard deviation of a sample must be a minimum around the sample mean.

$$s = \sqrt{\frac{\Sigma x^2}{N}}$$

Only occasionally will the sample mean and population mean coincide exactly. In all other cases, the variation around the sample mean will be smaller than the variation around the population mean.

Illustration. Suppose merely for illustrative purposes a population consisting of just six cases were split into two samples. The first sample contains values 2, 3, and 4, while the second contains values 0, 5, and 10.   The mean of the first sample ($\bar{X}_1$) is 3, and of the second sample ($\bar{X}_2$) 5. The population mean ($\mu$) is 4.

The standard deviation in the first sample

$$s_1 = \sqrt{\tfrac{2}{3}}$$
$$= .816$$

The standard deviation in the second sample

$$s_2 = \sqrt{\tfrac{50}{3}}$$
$$= 4.08$$

The standard deviation around $\mu$ in the entire universe is larger for each of the two samples. Taken around $\mu$, the two standard deviations are, respectively,

$$\sigma = \sqrt{\tfrac{5}{3}}$$
$$= 1.29$$

and

$$\sigma = \sqrt{\tfrac{53}{3}}$$
$$= 4.20$$

In each case, the sample standard deviation was less than the standard deviation for that sample when taken around the universe parameter. This is an example of the tendency for $s$ to be smaller than $\sigma$.

The sample standard deviation is therefore a *biased* estimate. For this statistic, the mean of the sampling distribution is not identical with the population parameter. By taking the standard deviation around the sample mean, which gives a minimum result, we have lost one *degree of freedom*. Degrees of freedom refer to the number of observations that are free to vary. This provides a basis for computing the standard error of the mean when the sample standard deviation must be used in place of $\sigma$.

The formula in Sec. 15-2 is modified to be

$$\sigma_{\bar{X}} = \frac{s}{\sqrt{N-1}}$$

The use of $N - 1$ in the denominator in place of $N$ compensates for the tendency of $s$ to underestimate $\sigma$. The decrease in standard error as $N$ increases is illustrated in Fig. 15-2.

**Illustration.** Over a period of 10 years, the mean vocabulary score of entering freshmen is 72, with a standard deviation of 7. This year's entering class has a mean of 75 for a group of 100 students. Is this group better on the vocabulary test than preceding classes?

$$\mu = 72 \qquad \bar{X} = 75 \qquad \sigma = 7 \qquad N = 100$$

*Null hypothesis:* This year's sample came from a population having a mean of 72 and a $\sigma$ of 7. The deviation of the sample mean is due to sampling variability.

Computation of $z$ score:

$$\sigma_{\bar{X}} = \frac{\sigma}{\sqrt{N}}$$
$$= \frac{7}{\sqrt{100}}$$
$$= .7$$
$$z = \frac{\bar{X} - \mu}{\sigma_{\bar{X}}}$$
$$= \frac{3}{.7}$$
$$= 4.29$$

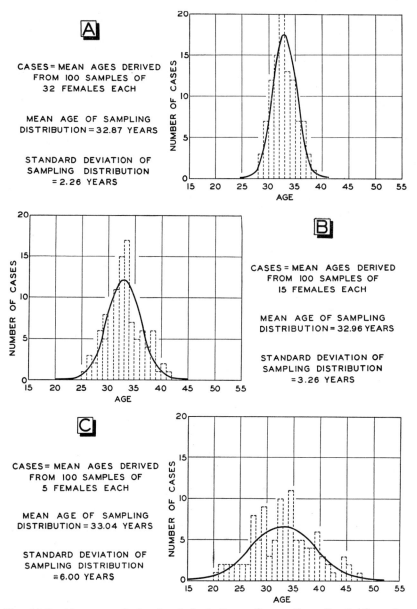

Fig. 15-2 Means and standard deviations of sampling distributions for samples of different sizes. It will be observed that the means of the sampling distribution tend to approximate the mean of the universe (in this instance, 32.98 years) but the standard deviations vary with size of the sample. The standard deviation decreases as the size of the sample is increased.

Probability that a sample this unusual would happen by chance:

From Table B, this would occur less than 1 time in 10,000. Since $z$ is greater than 1.96, this is significant at the .05 level. This year's class has done significantly better than its predecessors. There are many possible explanations for the improvement, and the statistical analysis only has demonstrated that there has been a significant improvement, not due to sampling variability.

**Illustration.** In a certain city the mean age of 50 convicted auto thieves is 19.2 years, while the mean age of all convicted criminals is 22.4. If the standard deviation of the age distribution of auto thieves is 2.1, does this indicate that auto thieves are younger than other criminals?

$$\mu = 22.4 \qquad \bar{X} = 19.2 \qquad s = 2.1 \qquad N = 50$$

*Null hypothesis:* Auto thieves have the same mean age as the mean age for all convicted criminals. The deviation of this sample mean arises from sampling variability.

Computation of $z$ score:

$$\sigma_{\bar{x}} = \frac{s}{\sqrt{N-1}}$$
$$= \frac{2.1}{\sqrt{49}}$$
$$= .3$$
$$z = \frac{\bar{X} - \mu}{\sigma_{\bar{x}}}$$
$$= \frac{19.2 - 22.4}{.3}$$
$$= \frac{-3.2}{.3}$$
$$= -10.67$$

Probability of a sample this unusual occurring by chance:

Since $z$ is greater than 1.96, the null hypothesis is rejected. Auto thieves are younger.

## 15-4    Student's $t$

We have stated previously that the sampling distribution of sample means will be approximately normal when there are 30 or more cases in the sample. In situations where the number of cases in the sample is less than 30, the sampling distribution is not close to normal, changing with the number of cases in the sample. With each $N$, the sampling distribution takes on a different shape. We are fortunate that a statistician, W. S. Gosset (better known under the pen name of "Student") has done the necessary computations for us.

"Student" uses $t$ as his measure of standard score, a substitute for $z$ where his sampling distributions are used. The computation of $t$ is exactly the same as the computation of $z$. Knowing $t$, we must go to the appropriate curve in order to determine the likelihood of so unusual a sample. The proper curve is found by using the number of degrees of freedom in the sample. Since 1 degree of freedom is lost through use of the sample mean, the appropriate number of degrees of freedom is $N - 1$.

Table C selects certain key values from a series of tables similar to Table B. Where Table B was based on the normal curve, each row of Table C uses a different sampling distribution. If the sample contains 11 cases, there are 10 degrees of freedom. Finding the tenth row in the first column, let us examine the first number in that row, 1.81. At the top of the column is .95. This means that 95 per cent of the cases are below this point. The number at the bottom of the column, .05, states that 5 per cent of the cases are above this standard score of 1.81. Using a two-tailed test of significance, a value as unusual as 1.81 would occur 10 per cent of the time.

Using both tails, the column labeled .975 and .025 gives the standard score necessary for significance at the .05 level with a specified number of degrees of freedom. With an infinite number of cases in a sample, the $t$ distribution becomes a normal distribution, as can be seen by the value of 1.96 in the row labeled infinity ($\infty$). The values of $t$ are so close to the values of $z$ necessary for significance that the normal curve is used for samples of 30 or more.

**Illustration.** A sample of 18 coal miners shows a mean income of $72 per week, compared with an average of $66 for all industrial workers covered in a census of a certain area. The standard deviation of income for coal miners is $4 per week. Is the mean income of coal miners higher than the mean income of all industrial workers?

$$\mu = 66 \qquad \bar{X} = 72 \qquad s = 4 \qquad N = 18$$

Computation of standard score:

$$\sigma_{\bar{X}} = \frac{s}{\sqrt{N-1}}$$
$$= \frac{4}{\sqrt{17}}$$
$$= \frac{4}{4.12}$$
$$= .97$$

$$t = \frac{\bar{X} - \mu}{\sigma_{\bar{X}}}$$
$$= \frac{72 - 66}{.97}$$
$$= \frac{6}{.97}$$
$$= 6.2$$

Probability of so unusual a sample occurring by chance:

$$\text{Degrees of freedom} = N - 1$$
$$= 18 - 1$$
$$= 17$$

At .05 level of significance, $t = 2.11$.

The value of $t = 6.2$ is larger than 2.11, so that samples this unusual would occur by chance less than 5 per cent of the time.

*Decision:* We would reject the null hypothesis. Coal miners on the average tend to receive higher wages than industrial workers generally.

## 15-5   Difference between Two Sample Means

In studying the sampling distribution of differences between two sample means when the samples are drawn independently of each other, the same principles apply that were discussed in Sec. 14-3. The population mean of the differences between sample means is 0. The *standard error of the difference between means,* $\sigma_{\bar{X}_1 - \bar{X}_2}$, is again equal to the square root of the sum of the squares of the standard errors.

$$\sigma_{\bar{X}_1 - \bar{X}_2} = \sqrt{\sigma_{\bar{X}_1}^2 + \sigma_{\bar{X}_2}^2}$$

In finding the standard error of differences between proportions, the difficulty is not knowing $p_u$. The population mean $\mu$ need not be known or estimated in working with the standard error of means. The estimate that is needed is the standard deviation of the universe $\sigma$. When working with $s$, the standard deviation of the sample, we lose 1 degree of freedom. Therefore, we lose 2 degrees of freedom when working with the difference between two samples, one for each sample standard deviation. The degrees of freedom is then $N_1 + N_2 - 2$. If the number of degrees of freedom is 30 or over, the normal curve is used as a good approximation. If the degrees of freedom are less than 30, Table C is used.

**Illustration.** The mean productive efficiency of a high-morale group is predicted to be higher than a low-morale group on the basis of existing theories of industrial relations. A sample of 26 high-morale workers at a large plant has an efficiency score of 81 with a standard deviation of 6. At the same plant, a low-morale group of 61 workers has an efficiency score of 76 and a standard deviation of 5. Are we justified in inferring that these data support existing theories of industrial relations?

In this analysis, we shall use a one-tailed test of significance, since the direction of differences expected was stated before the samples were observed. We shall see if the standard score is beyond the point which has 5 per cent of the cases above it, rather than 2.5 per cent of the cases.

*Null hypothesis:* These two samples come from populations having the same mean efficiency. The observed differences between sample means are due to sampling variability.

$$\bar{X}_1 = 81 \qquad s_1 = 6 \qquad N_1 = 26 \qquad \bar{X}_2 = 76 \qquad s_2 = 5 \qquad N_2 = 61$$

Computations of standard score:

$$\sigma_{\bar{X}_1} = \frac{s_1}{\sqrt{N_1 - 1}}$$
$$= \frac{6}{\sqrt{26 - 1}}$$
$$= \frac{6}{5}$$
$$= 1.20$$

$$\sigma_{\bar{X}_2} = \frac{s_2}{\sqrt{N_2 - 1}}$$
$$= \frac{5}{\sqrt{61 - 1}}$$
$$= \frac{5}{7.75}$$
$$= .65$$

$$\sigma_{\bar{X}_1 - \bar{X}_2} = \sqrt{\sigma_{\bar{X}_1}{}^2 + \sigma_{\bar{X}_2}{}^2}$$
$$= \sqrt{(1.20)^2 + (.65)^2}$$
$$= \sqrt{1.4400 + .4225}$$
$$= \sqrt{1.8625}$$
$$= 1.36$$
$$t = \frac{(\bar{X}_1 - \bar{X}_2) - 0}{\sigma_{\bar{X}_1 - \bar{X}_2}}$$
$$= \frac{81 - 76}{1.36}$$
$$= \frac{5}{1.36}$$
$$= 3.7$$

In order to determine how often a difference this great in this direction would occur through sampling variability, the following computation is made:

$$\text{Degrees of freedom} = N_1 + N_2 - 2$$
$$= 26 + 61 - 2$$
$$= 85$$

Since the number of degrees of freedom is over 30, we can use the normal curve. From Table B, a $z$ score of 1.645 or more would occur 5 per cent of the time or less. This $z$ score is 3.7, so that the observed differences are significant at the .05 level.

*Decision:* We reject the null hypothesis. The sample results bear out our expectations.

## 15-6    Confidence Limits of Means

Finding the 95 per cent confidence limits when we know the sample mean involves no different logic than in the case of proportions. Either 1.96 or the appropriate $t$ is multiplied by the standard error of means to obtain the maximum deviation of sample mean from population mean. Figures 15-3 and 15-4 show in generalized form the 95 per cent and 99 per cent confidence limits.

**Illustration.** In a sample of 25 families handled by a welfare agency, it was found that the mean number of years they received aid was 1.9, with a standard deviation of .8. What are the 95 per cent confidence limits for the mean number of years of aid received for all the cases handled by the agency?

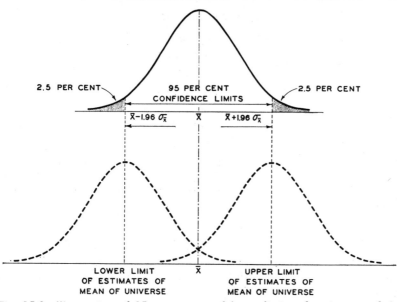

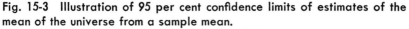

Fig. 15-3  Illustration of 95 per cent confidence limits of estimates of the mean of the universe from a sample mean.

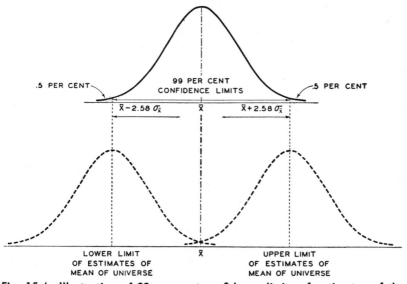

Fig. 15-4  Illustration of 99 per cent confidence limits of estimates of the mean of the universe from a sample mean.

Computation of standard error of means:

$$\sigma_{\bar{x}} = \frac{s}{\sqrt{N-1}}$$

$$= \frac{.8}{\sqrt{25-1}}$$

$$= \frac{.8}{4.9}$$

$$= .16$$

Computation of standard score:

$$\text{Degrees of freedom} = N - 1$$
$$= 25 - 1$$
$$= 24$$

Since the number of degrees of freedom is less than 30, we must use the $t$ distribution rather than the normal curve.

For a two-tailed test with 24 degrees of freedom, the $t$ is equal to 2.06 at the .05 level of significance.

The deviation of the sample mean from the population mean must not exceed 2.06 (.16) = .33.

Computation of confidence limits:

The lower limit is 1.9 − .33 = 1.57 years.

The upper limit is 1.9 + .33 = 2.23 years.

At the 95 per cent level of confidence, the population mean is probably between 1.57 and 2.23 years.

## 15-7   Summary

The sampling distribution of sample means is approximately normal if the number of cases in the sample is 30 or over.

If the number of cases in the sample is under 30, Student's $t$ distribution is used with $N - 1$ degrees of freedom.

$$\sigma_{\bar{x}} = \frac{\sigma}{\sqrt{N}} = \frac{s}{\sqrt{N-1}}$$

The sampling distribution of differences between means tends to be normal if the degrees of freedom, equal to $N_1 + N_2 - 2$, is 30 or over.

$$\sigma_{\bar{x}_1 - \bar{x}_2} = \sqrt{\sigma_{\bar{x}_1}^2 + \sigma_{\bar{x}_2}^2}$$

If the degrees of freedom is under 30, Student's $t$ distribution is used for the difference between sample means.

### PROBLEMS FOR CHAP. 15

**1.** A paper-and-pencil test has been devised to measure psychopathic tendencies. It has been standardized on youths aged 18 to 24, giving a mean of 17 and standard deviation of 3. A random sample of 65 youths in the same age group at a state reformatory has a mean of 18 and standard deviation of 4.5. Is this criminal group in tendency toward psychopathy significantly different than the population on which the test was standardized?                *Ans.* Yes

**2.** The mean score on a test of social adjustment for high-school pupils is 50, with a standard deviation of 10. A sample of 19 students who were toilet-trained at a relatively late age has an adjustment mean of 54. Is the difference significant?                *Ans.* No

**3.** A sample of 400 persons in 1948 and another sample of 400 in 1955 were interviewed as to their conception of America's most pressing problems. The 1948 sample mentioned problems connected with foreign affairs often enough to make the mean number of mentions of foreign relations 2.4, with a standard deviation of .6. The 1955 sample mentioned foreign affairs problems only to the extent of a mean of 2.0 times, with a standard deviation of .4. Is there a significant decrease between the two samples in concern about foreign relations?

*Ans.* Yes

**4.** A random sample of 400 graduates of Syracuse University reported a mean number of jobs since graduation of 3.4, with a standard deviation of .7. What are the 95 per cent confidence limits?

*Ans.* 3.33 to 3.47

# 16 Prediction and Regression

This chapter is concerned with finding the relationship between two variables. It describes an approach to the problem of discerning what happens to one variable when the other variable increases or decreases. A common practice is to conceive of the problem as attempting to predict values in one series on the basis of known values in some other series. As used in this context, "predict" does not necessarily imply forecasting some future sequence but merely means deriving the "best possible estimate." This chapter is devoted to a discussion of techniques used in the study of predictive relationships of two variables where information is available for all the cases in each series.

## 16-1   Prediction and the Relationship between Variables

Suppose that a group of men was measured with respect to two variables, the number of years of schooling each had completed and the amount of income each receives. If we could describe the relationship between education and income, it would be of considerable help in calculating predictive estimates for these two variables. If the two variables were closely related, then knowledge of a man's education would enable us to make a

170

relatively close estimate of the amount of his current income. Or, similarly, if we knew a man's income, we could make a reasonable estimate of his educational background.

Table 16-1 presents years of schooling and income for 24 men approximately 40 years of age. Let us attempt to predict (or estimate) income from a knowledge of years of schooling. In such a situation income is usually considered dependent on education. Income, therefore, is called the *dependent variable*

Table 16-1   Relationship between Education (Number of School Years Completed) and Income for a Sample of 24 Males Approximately 40 Years of Age

| (1)<br><br>Case<br>designation | (2)<br><br>Education,<br>school years completed<br><br>($X$) | (3)<br><br>Income,<br>thousands of dollars<br><br>($Y$) |
|:---:|:---:|:---:|
| A | 6 | 3 |
| B | 6 | 5 |
| C | 7 | 4 |
| D | 7 | 4 |
| E | 7 | 3 |
| F | 8 | 4 |
| G | 8 | 6 |
| H | 8 | 5 |
| I | 9 | 3 |
| J | 9 | 4 |
| K | 10 | 4 |
| L | 11 | 5 |
| M | 11 | 7 |
| N | 12 | 5 |
| O | 12 | 5 |
| P | 12 | 8 |
| Q | 13 | 7 |
| R | 13 | 5 |
| S | 13 | 9 |
| T | 14 | 11 |
| U | 15 | 8 |
| V | 15 | 6 |
| W | 16 | 7 |
| X | 16 | 6 |

($Y$), since it is treated as dependent on education, with schooling used to estimate income. In this problem education is referred to as the *independent variable* ($X$).

## 16-2   Scatter Diagram

The first step in this analysis is to plot every case on a coordinate grid. The plotting of all the points on a chart of this kind produces a *scatter diagram*. The scatter diagram provides

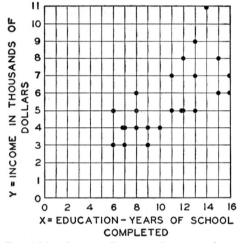

Fig. 16-1   Scatter diagram showing educational status and income of 24 men approximately 40 years of age.

immediate knowledge of the approximate nature of the relationship between the two series. Figure 16-1 indicates that there is a definite positive relationship between the two variables; that is, when education increases, income tends to increase.

Another reason for constructing a scatter diagram is to determine whether or not it is proper to use a straight line for expressing the relationship between the two variables. In this course, we shall be concerned only with straight-line or rectilinear relationships, but we always must examine the scatter diagram to be sure our methods are appropriate. If a nonlinear (curvilinear) relationship were found, then we would be obliged to use some other method of analysis. The constancy of the ratio of change of the two variables determines whether the relationship is rectilinear or curvilinear. If the amount of change of the two variables bears a constant ratio, the relationship is rectilinear, but if it does not, then the relationship is curvilinear. A unit change in $X$ produces a constant change in $Y$ when there is a straight-line relationship; $Y$ always increases (or decreases) the same amount when $X$ increases 1 unit.

## 16-3  Least-squares Line

Suppose nothing were known about the relationship between education and income and it were necessary to predict the incomes of these 24 men. What would be the best estimates? One possibility would be to select the modal income and predict the mode for every case. This would give six correct guesses, more than if any other number were continually guessed. Another possibility would be to choose for each case the mean of the entire series (5.58). If this were done, not a single case would correspond exactly to this figure, but the sum of the squared deviations from the estimates would be a minimum (Sec. 5-5). The determination of whether the mode or the mean is a wiser choice depends entirely on the criterion by which success is judged.

We shall set up our criterion for good prediction in advance when we attempt to predict on the basis of the observed relationship between income and education. Our prediction shall be based on the *least-squares line*. It is the line for which the squared vertical deviations of the points from the line are a minimum. The concept will be clarified considerably after we derive a mathematical statement of the least-squares line and plot the line from this equation on the scatter diagram of income and education.

Every straight line can be written in the form

$$Y = a + bX$$

As usual, $a$ and $b$ represent constants, but here they have a special meaning. The $a$ is the point at which the line crosses the $Y$ axis, called the $Y$ *intercept*. The $b$ stands for the *slope* of

**Table 16-2  Calculation of Least-squares Line** (Data from Table 16-1, Education and Income for 24 Adult Males)

| (1) Case designation | (2) Education, school years completed ($X$) | (3) Income, thousands of dollars ($Y$) | (4) $(X)(Y)$ | (5) $X^2$ | (6) $Y^2$ |
|---|---|---|---|---|---|
| A | 6 | 3 | 18 | 36 | 9 |
| B | 6 | 5 | 30 | 36 | 25 |
| C | 7 | 4 | 28 | 49 | 16 |
| D | 7 | 4 | 28 | 49 | 16 |
| E | 7 | 3 | 21 | 49 | 9 |
| F | 8 | 4 | 32 | 64 | 16 |
| G | 8 | 6 | 48 | 64 | 36 |
| H | 8 | 5 | 40 | 64 | 25 |
| I | 9 | 3 | 27 | 81 | 9 |
| J | 9 | 4 | 36 | 81 | 16 |
| K | 10 | 4 | 40 | 100 | 16 |
| L | 11 | 5 | 55 | 121 | 25 |
| M | 11 | 7 | 77 | 121 | 49 |
| N | 12 | 5 | 60 | 144 | 25 |
| O | 12 | 5 | 60 | 144 | 25 |
| P | 12 | 8 | 96 | 144 | 64 |
| Q | 13 | 7 | 91 | 169 | 49 |
| R | 13 | 5 | 65 | 169 | 25 |
| S | 13 | 9 | 117 | 169 | 81 |
| T | 14 | 11 | 154 | 196 | 121 |
| U | 15 | 8 | 120 | 225 | 64 |
| V | 15 | 6 | 90 | 225 | 36 |
| W | 16 | 7 | 112 | 256 | 49 |
| X | 16 | 6 | 96 | 256 | 36 |
| Total | 258 | 134 | 1,541 | 3,012 | 842 |

$$b = \frac{N\Sigma XY - (\Sigma X)(\Sigma Y)}{N\Sigma X^2 - (\Sigma X)^2}$$

$$= \frac{24(1,541) - (258)(134)}{24(3,012) - (258)^2}$$

$$= \frac{36,984 - 34,572}{72,288 - 66,564}$$

$$= \frac{2,412}{5,724}$$

$$= .42$$

$$a = \frac{\Sigma Y - b\Sigma X}{N}$$

$$= \frac{134 - .42(258)}{24}$$

$$= \frac{134 - 108.36}{24}$$

$$= 1.07$$

the line, the amount that $Y$ increases or decreases when $X$ increases one unit.

The main problem is to find the $a$ and the $b$ that will give the least-squares line for these data. By slightly more advanced mathematics, the two constants for the least-squares line can be found quickly by means of the following formulas:

$$b = \frac{N\Sigma XY - (\Sigma X)(\Sigma Y)}{N\Sigma X^2 - (\Sigma X)^2}$$

$$a = \frac{\Sigma Y - b\Sigma X}{N}$$

**Illustration.** Table 16-2 presents the various steps in computing a least-squares line.

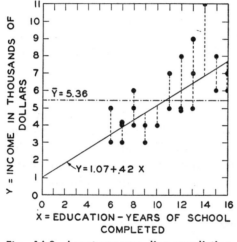

Fig. 16-2   Least-squares line predicting income from education with vertical deviations of observations from predictions.

This least-squares line, $Y = 1.07 + .42X$, is drawn in Fig. 16-2, along with the vertical deviations from the least-squares line. To plot a least-squares line, two values are substituted for $X$, and the two corresponding $Y$ values are obtained. For example, when $X$ is 10, $Y$ is 5.27. These values (10 and 5.3) are plotted on the scatter diagram as a single point. By plotting another point, a straight line can be drawn connecting the two points.

## 16-4   Regression Lines

The least-squares line predicting $Y$ from $X$ is also known as the *regression line* of $Y$ on $X$. This line keeps the squared vertical deviations at a minimum, so that the variance of the observations around the line must also be a minimum. Predictions derived from the least-squares line are an improvement over predictions in which mean income is used each time as the prediction. If a man has had 8 years of schooling, then the least-squares line predicts that his income is

$$Y = 1.07 + .42(8)$$
$$= 1.07 + 3.36$$
$$= 4.4 \text{ thousand dollars}$$

**Problem.** Find the linear regression coefficients for estimating $X$ when knowing $Y$, predicting education when income is known. *Hint:* In the formulas for $a$ and $b$, substitute $X$ for $Y$, and vice versa.

*Ans.* $a = 4.8$; $b = 1.07$

If the regression line of $X$ on $Y$ in the problem above is drawn in Fig. 16-2, it will be noted that the two regression lines cross. The point at which they intersect always has the value $(\bar{X}, \bar{Y})$. In this particular case the point will have the value (10.75, 5.58). If the lines do not meet exactly at that point, the discrepancy may be due to inaccuracies in the scatter diagram.

## 16-5   Summary

The dependent variable is the variable to be predicted.

The independent variable provides the basis for prediction.

The variance of the observations is a minimum around the least-squares line.

$Y = a + bX$, where $a$ is the $Y$ intercept and $b$ is the slope.

$$b = \frac{N\Sigma XY - (\Sigma X)(\Sigma Y)}{N\Sigma X^2 - (\Sigma X)^2}.$$

$$a = \frac{\Sigma Y - b\Sigma X}{N}.$$

The two regression lines always cross at the point $(\bar{X}, \bar{Y})$.

## PROBLEM FOR CHAP. 16

1. A sample of females 6 years of age and older were asked how many times per month they attend religious service. The data showing the relationship between age and religious attendance are presented below. Draw a scattergram and attempt to predict the position of the least-squares line. Calculate the formula of the least-squares line.

| Age (X) | Number of services (Y) | Age (X) | Number of services (Y) |
|---------|------------------------|---------|------------------------|
| 6       | 2                      | 24      | 1                      |
| 6       | 4                      | 24      | 0                      |
| 7       | 4                      | 26      | 1                      |
| 8       | 3                      | 29      | 2                      |
| 10      | 0                      | 31      | 4                      |
| 10      | 3                      | 35      | 0                      |
| 13      | 6                      | 39      | 2                      |
| 14      | 8                      | 42      | 4                      |
| 17      | 4                      | 43      | 9                      |
| 18      | 2                      | 48      | 6                      |
| 20      | 0                      | 51      | 4                      |

*Ans.* $Y = 2.41 + .03X$

# 17 Linear Correlation

In the preceding chapter, it was pointed out how the least-squares line can be used for prediction on the basis of an observed relationship between two quantitative variables. We are concerned here with developing a measure of the degree (amount) of observed relationship and the corresponding ability to predict.

## 17-1 Perfect Correlation

Figure 17-1 shows scattergrams for two sets of variables which are very closely related. Indeed, the degree of association is so high that the least-squares line passes through every observed case. This indicates perfect ability to predict (or estimate) values of one variable which correspond to the known values of the other variable.

It will be observed from Fig. 17-1$A$ that a high value of $X$ is always associated with a high value of $Y$. Correspondingly, a low value of $X$ indicates a low $Y$ value. (These variables are said to show a positive relationship or correlation.) Figure 17-1$B$ is dif-

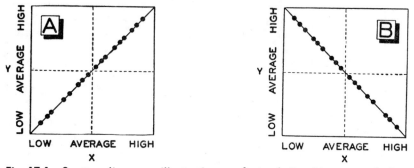

Fig. 17-1   Scatter diagrams illustrating perfect relationship or correlation. (A) perfect positive correlation; (B) perfect negative correlation.

ferent from Fig. 17-1A in that it shows the opposite type of relationship. Prediction is just as perfect, and the association is just as pronounced (strong). In Fig. 17-1B high values on the X axis are associated with low values on the Y axis; and similarly, low X values occur together with high Y values. This is known as negative (inverse) relationship or correlation.

Figure 17-1A illustrates *perfect positive correlation*. This is an illustration of perfect correlation because the least-squares line gives perfect prediction. The term "perfect" as used in this context implies merely certainty in prediction.

On the other hand, Fig. 17-1B portrays *perfect negative correlation*. High values in one variable occur together with low values in the other variable. Prediction is again absolutely exact.

An example of positive relationship or correlation can be observed between income and education. People of high income tend to have relatively more education, while people of less education have, in general, low incomes. On the other hand, there is a negative relationship between income and the birth rate. People of higher incomes tend to have fewer children.

Almost never in the social sciences is it possible to find two variables that are perfectly correlated. In studying the relationship of two variables, we are consciously ignoring the effect of other variables. The influence of these other variables usually reduces our ability to predict. Figure 17-2 illustrates moderate positive relationship and moderate negative relationship. The least-squares lines give good, but not perfect, prediction.

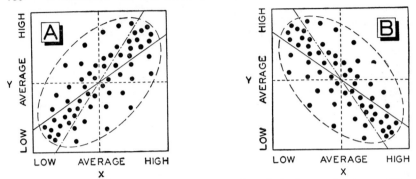

**Fig. 17-2  Scatter diagrams showing:** (*A*) relatively moderate positive correlation and (*B*) relatively moderate negative correlation.

## 17-2  No Correlation

When there is no relationship at all between two variables, the least-squares lines do not improve our ability to predict. Know-

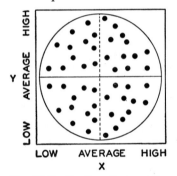

**Fig. 17-3**  Scatter diagram in which there is no correlation between the two variables.

ing the value of $Y$ for any case does not help us to predict $X$, and the knowledge of $X$ tells us nothing about $Y$. The criterion specified for accuracy of prediction is to minimize the sum of the squared deviations (Sec. 16-3). Lacking any further aid, the best prediction for each case of $Y$ is the mean of the $Y$ values, and similarly, the best prediction for each case of $X$ is the mean of the $X$ values. This is true because the sum of squared deviations around the mean is a minimum (Sec. 5-5). Figure 17-3 shows two least-squares lines when there is no relationship between $X$ and $Y$.

## 17-3  Covariation

In the preceding sections high and low values of $X$ and $Y$ were discussed. A high value is one considerably above the mean, while a low value is one considerably below the mean. In attempting to derive a single number to express the ideas portrayed in scatter diagrams, each case can be symbolized in terms of its deviation from $\bar{X}$ and $\bar{Y}$. These deviations can be written as $x$ and $y$. Each case will have both an $x$ and a $y$, since every case has been measured for both variables.

If the two deviations for every case were multiplied together and the products added, the result is symbolized as $\Sigma xy$, the *covariation*. The covariation provides a simple method of determining whether the relationship between two variables is positive, negative, or zero.

Let us examine what happens to the covariation when a relationship is positive. High values of two variables tend to be found in the same cases, and similarly low values tend to be found together. The $x$ and $y$ for high-value cases are both positive, and the low-value cases have negative numbers for both $x$ and $y$. The product of two positive numbers is positive, and the result of multiplying two negative numbers together is also positive (Supplementary Explanation in Sec. 5-5). The covariation in a positive relationship will tend to be positive.

When there is a negative relationship, high values of $X$ will be associated with low values of $Y$, and vice versa. This means a positive deviation in $X$ usually will be multiplied by a negative deviation in $Y$. Correspondingly, a negative $x$ will be multiplied by a positive $y$ in most instances. The result of both these processes is negative, so that $\Sigma xy$ will be negative when there is a negative association between the variables.

When there is little or no relationship between the variables, a positive $x$ is likely to be found equally with a positive $y$ or a negative $y$. In the first instance, the product would be positive; in the second, negative. The same would be found true for low values of $x$. There would be no tendency to obtain either positive or negative results. $\Sigma xy$ would be zero, since the minus signs and plus signs would balance exactly.

To sum up, the size, along with the sign of the covariation, indicates the strength and direction of the relationship between two variables.

## 17-4  Deficiency of Covariation

Unfortunately, covariation possesses a serious limitation which precludes its use as a measure for determining the degree of relationship between two variables. It is affected by the units used in the measurement of both $X$ and $Y$. If $X$, for example, is expressed in very large numbers, then $x$ will also be expressed in large numbers. Therefore, $\Sigma xy$ would tend to be large.

To overcome this deficiency, the covariation is divided by a number which is influenced by the units used in exactly the same way as the covariation. The number selected as a divisor must be equally affected by both the units of $X$ and the units of $Y$. Accordingly, the geometric mean of the sums of squared deviations for $X$ and $Y$ is used. This is written, as in Sec. 5-15,

$$\sqrt{(\Sigma x^2)(\Sigma y^2)}$$

## 17-5   Coefficient of Correlation

The *coefficient of correlation* $(r)$ is defined in terms of the preceding section. It, therefore, will provide a measure of the strength and direction of association which can be used for direct comparisons, since the units of analysis do not affect its value.

$$r = \frac{\Sigma xy}{\sqrt{(\Sigma x^2)(\Sigma y^2)}}$$

Since the covariation is in the numerator, the coefficient of correlation can be either negative or positive. The value of $r$ ranges from $-1$ to $+1$. If $r$ is $-1$, there is a perfect negative relationship between the two variables, whereas if $r$ is $+1$ there is a perfect positive relationship. In either case, prediction on the basis of the least-squares line would be perfect. If there is no relationship between two variables, knowledge of the value of a case for one variable does not aid in predicting its value for the other variable. Accordingly, $r$ is 0.

The coefficient of correlation is a measure of the ability to predict on the basis of the least-squares regression line. If a curvilinear relationship exists between two variables, $r$ will tend to be low depending on the degree of curvilinearity that exists. Since $r$ measures the predictive ability of the least-squares line, a straight line hardly can be a good predictor when the relationship is shown best by a curve. In Fig. 17-4, there is a very strong relationship between $X$ and $Y$, but the least-squares line is a poor predictor. The $r$ would be close to 0. More advanced courses in statistics usually include methods for handling curvilinear relationships. By plotting data on a scattergram, it is possible to ascertain whether the relationship is rectilinear or curvilinear. The coefficient of correlation $(r)$ is usually applicable only in situations where the relationship between the two variables can be adequately expressed by a straight line.

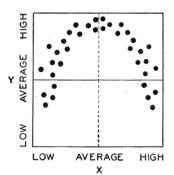

Fig. 17-4  A scatter diagram showing a relatively high curvilinear relationship. An "$r$" computed for a distribution of this kind would be very close to zero.

If the least-squares line gives perfect prediction, then $r$ will be $+1$ or $-1$. An $r$ of 0 will mean that there is no relationship between the two variables. Seldom, if ever, in actual practice do we obtain perfect correlation or absolutely no correlation. Coefficients of correlation are usually some number between 0 and 1, either positive or negative.

## 17-6  Total Variation

If, for example, an $r$ of $-.6$ were found, the minus sign shows that the relationship is negative. The meaning of the .6 is not so clear. A relationship of this amount can be used for predictive purposes, but prediction would not be perfect. In order to understand for prediction purposes the size of $r$, let us assume a situation where no relationship exists between two variables. If an attempt were made to predict the value of $Y$ when the value of $X$ for each case is known, the best prediction where $r$ is 0 is $\bar{Y}$. The mean of the $Y$ values will minimize the sum of the squared deviations (Sec. 17-2). This provides a base against which to measure our ability to predict with the least-squares line. The variation around $\bar{Y}$ is the error in prediction when there is no relationship between the variables. The variation around the line at the mean is referred to as the *total variation*. Figure 17-5A

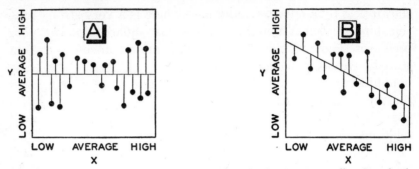

Fig. 17-5  (A) When no correlation exists, the least-squares line is a horizontal line at $\bar{Y}$, and the unexplained variation is equal to the total variation. (B) Moderate negative correlation with least-squares line indicating that the unexplained variation is less than the total variation.

shows the vertical deviations from the line at $\bar{Y}$ which, squared and summed, form the total variation.

## 17-7  Unexplained Variation

If there is a relationship between the two variables, the least-squares line will not be horizontal at $\bar{Y}$. It will slope up or down, giving increased ability to predict. The errors of prediction, shown in Fig. 17-5B, will produce variation around the least-squares line. This variation around the predictive line will be called the *unexplained variation*. It is the variation remaining after we take into account the relationship between $X$ and $Y$. The unexplained variation is the error of prediction from the regression line.

## 17-8  Explained Variation

The unexplained variation can never be greater than the total variation. Where there is no relationship between the two variables, the least-squares line is a horizontal line at $\bar{Y}$. In such a case, the two lines coincide, and the unexplained variation is equal to the total variation. Where there is some increased ability to predict from a least-squares regression line, the variation around the regression line is less than the variation around

$\bar{Y}$. The improvement in prediction is measured by the *explained variation*. It is found by subtracting the unexplained variation from the total variation. If there is perfect prediction, the explained variation equals the total variation, since there is no unexplained variation. When there is no relationship, the explained variation is 0.

## 17-9 Coefficient of Determination

We now have reached a point where the value of the correlation coefficient takes on additional meaning. The *coefficient of determination* $r^2$ indicates the proportion of the total variation that is explained by the least-squares line. If $r$ is $-.6$, then $r^2$ is .36. Therefore, 36 per cent of the total variation is explained by the least-squares line. Our predictions show a 36 per cent improvement compared to a situation where there is no relation between the two variables. The greater the improvement in prediction, the stronger must be the relationship between the variables. Therefore, $r^2$ is a measure of the strength of the observed relationship.

The discussion in the foregoing sections can be expressed by the following formulas:

$$r^2 = \frac{\text{explained variation}}{\text{total variation}}$$

Another way of writing the same idea is

$$r^2 = 1 - \frac{\text{unexplained variation}}{\text{total variation}}$$

Since the correlation coefficient $r$ is the square root of the coefficient of determination $r^2$,

$$r = \sqrt{\frac{\text{explained variation}}{\text{total variation}}}$$

$$r = \sqrt{1 - \frac{\text{unexplained variation}}{\text{total variation}}}$$

**Problems. 1.** If the explained variation is equal to the total variation, what is $r$? *Ans.* $-1$ or 1

**2.** Find the coefficient of determination when the unexplained variation equals the total variation.                                      *Ans.* 0

**3.** Find $r$ when the total variation is 64 and the unexplained variation is 32.                                      *Ans.* .707 or $-.707$

**4.** When $r$ is .4, what proportion of the total variation is explained?
                                      *Ans.* 16 per cent

**5.** When the coefficient of determination is .81, what is the correlation coefficient?                                      *Ans.* .9 or $-.9$

**6.** If $r$ is .4 and the total variation is 50, what is the unexplained variation?                                      *Ans.* 42

We now have developed a measure of the strength of the relationship between two variables, the correlation coefficient. It is a pure number, independent of the units of measurement for either of the variables. The square of the correlation coefficient, the coefficient of determination, gives the proportion of the total variance that can be predicted by knowledge of the relationship. The stronger the relationship, the better the prediction, so that $r$ is a measure of both strength of relationship and of prediction.

### 17-10   Short Method for Computing $r$ from Ungrouped Data

In Sec. 17-5 the correlation coefficient is defined as

$$r = \frac{\Sigma xy}{\sqrt{(\Sigma x^2)(\Sigma y^2)}}$$

That formula is obviously inadequate for computational purposes. The decimals involved in squaring deviations from the mean make the process time-consuming (Sec. 6-10). To save effort, we shall derive by the rules of summation a formula for the correlation coefficient that does not use the deviations from the mean. This computational formula for the correlation coefficient is

$$r = \frac{\Sigma XY - \dfrac{(\Sigma X)(\Sigma Y)}{N}}{\sqrt{\left[\Sigma X^2 - \dfrac{(\Sigma X)^2}{N}\right]\left[\Sigma Y^2 - \dfrac{(\Sigma Y)^2}{N}\right]}}$$

**Proof.** To prove that the formula above is equivalent to the definition of the correlation coefficient given in (Sec. 17-5),

$$r = \frac{\Sigma xy}{\sqrt{(\Sigma x^2)(\Sigma y^2)}}$$

In the denominator, under the square-root sign, are the variations in both $X$ and $Y$. Variation is equal to $N$ times the variance (Sec. 6-10). In Sec. 6-10, it was proved that

$$\sigma^2 = \frac{\Sigma X^2}{N} - \left(\frac{\Sigma X}{N}\right)^2$$

Multiplying by $N$ will give a formula for variation.

$$\Sigma x^2 = N\sigma^2 = N\left[\frac{\Sigma X^2}{N} - \left(\frac{\Sigma X}{N}\right)^2\right]$$

Carrying out the multiplication by $N$ will make the first term $\Sigma X^2$. The second term, $(\Sigma X/N)^2$, can be written as $(\Sigma X)^2/N^2$. If this is multiplied by $N$, then $N$ will be in the numerator and $N^2$ in the denominator. The result is the same as just dividing by $N$. Therefore,

$$\Sigma x^2 = \Sigma X^2 - \frac{(\Sigma X)^2}{N}$$

Obviously, the same formula could be used to express variation in the $Y$ variable. In the original formula at the beginning of this section, the derived formulas can be substituted for variation in $X$ and $Y$. This gives

$$r = \frac{\Sigma xy}{\sqrt{\left[\Sigma X^2 - \frac{(\Sigma X)^2}{N}\right]\left[\Sigma Y^2 - \frac{(\Sigma Y)^2}{N}\right]}}$$

The remaining problem is to derive a formula for the covariation in the numerator. By definition,

$$\Sigma xy = \Sigma[(X - \bar{X})(Y - \bar{Y})]$$

Let us multiply $(X - \bar{X})$ by $(Y - \bar{Y})$.

$$
\begin{array}{r}
X - \bar{X} \\
Y - \bar{Y} \\
\hline
XY - Y\bar{X} \\
- X\bar{Y} + \overline{XY} \\
\hline
XY - Y\bar{X} - X\bar{Y} + \overline{XY}
\end{array}
$$

Substituting,

$$\Sigma xy = \Sigma(XY - Y\bar{X} - X\bar{Y} + \overline{XY})$$

Summating each of the terms, and recalling that $\bar{X}$ and $\bar{Y}$ are constants,

$$\Sigma xy = \Sigma XY - \Sigma Y\bar{X} - \Sigma X\bar{Y} + \Sigma \overline{XY}$$
$$\Sigma xy = \Sigma XY - \bar{X}\Sigma Y - \bar{Y}\Sigma X + N\overline{XY}$$

Wherever $\bar{X}$ and $\bar{Y}$ appear, let us insert their defining formulas.

$$\Sigma xy = \Sigma XY - \frac{\Sigma X}{N}(\Sigma Y) - \frac{\Sigma Y}{N}(\Sigma X) + N\left(\frac{\Sigma X}{N}\right)\left(\frac{\Sigma Y}{N}\right)$$

In the fourth term, $N$ appears in the denominator two times. This is the same as dividing by $N^2$. At the same time, $N$ appears in the numerator. Multiplying by $N$ and dividing by $N^2$ is equivalent to dividing by $N$.

$$\Sigma xy = \Sigma XY - \frac{(\Sigma X)(\Sigma Y)}{N} - \frac{(\Sigma Y)(\Sigma X)}{N} + \frac{(\Sigma X)(\Sigma Y)}{N}$$

The numerator of the second, third, and fourth terms is $(\Sigma X)(\Sigma Y)$. The denominator is $N$ for each of those terms. Therefore, the three terms are exactly equal to each other. We subtract the second and third terms and add the fourth, so that the result is the same as subtracting the term once.

$$\Sigma xy = \Sigma XY - \frac{(\Sigma X)(\Sigma Y)}{N}$$

Returning to the original defining equation for $r$, and making all the substitutions,

$$r = \frac{\Sigma XY - \dfrac{(\Sigma X)(\Sigma Y)}{N}}{\sqrt{\left[\Sigma X^2 - \dfrac{(\Sigma X)^2}{N}\right]\left[\Sigma Y^2 - \dfrac{(\Sigma Y)^2}{N}\right]}}$$

**Illustration.** Now that we have derived a method of computing the correlation coefficient without finding deviations from the mean, let us apply it to the data used in Chap. 16. Table 16-2 gives much of the data required for substitution in the formula above.

$$r = \frac{1{,}541 - \dfrac{(258)(134)}{24}}{\sqrt{\left[3{,}012 - \dfrac{(258)^2}{24}\right]\left[842 - \dfrac{(134)^2}{24}\right]}}$$

$$r = \frac{1{,}541 - 1{,}440.5}{\sqrt{(3{,}012 - 2{,}773.5)(842 - 748.2)}}$$

$$r = \frac{100.5}{\sqrt{(238.5)(93.8)}}$$

$$r = \frac{100.5}{\sqrt{22{,}371.3}}$$

$$r = \frac{100.5}{149.6}$$

$$r = .67$$

The preceding illustration shows a positive relation between education and income for 24 adult males. The size of the correlation coefficient, .67, indicates that the proportion of the total variance explained by the relationship between the variables is $(.67)^2$, or .45.

## 17-11    Correlation and Slope of Regression Lines

If there had been no relationship between $X$ and $Y$ in the preceding example, the least-squares lines would have been at $\bar{X}$ and $\bar{Y}$ (Fig. 17-3). But there was an observed relationship between the two variables. When $X$ increases, $Y$ increases, and when $Y$ increases, $X$ increases. Accordingly, there is a positive slope for both regression lines. When there is no relationship, the slope of the regression lines is 0. $Y$ does not increase nor decrease when $X$ changes, but rather remains constant as $\bar{Y}$. When there is some relationship, the regression lines have either a positive or negative slope.

The greater the slope, the stronger the relationship between the variables. It is difficult to use the slope of a regression line as a measure of relationship since it changes with the units of $X$ and $Y$. The slope for one least-squares line, $b_{XY}$, is the change in $Y$ per unit change in $X$. The other least-squares line has a slope, $b_{YX}$, in the terms of the change in $X$ per unit change in $Y$. The units in which the two slopes are measured are, therefore, the exact opposite of each other.

The multiplication together of the two slopes will produce a pure number, not affected by the units of $X$ or $Y$. The geometric mean of the two slopes is the correlation coefficient (Sec. 5-10).

$$r = \sqrt{b_{XY}b_{YX}}$$

Using the same example from Secs. 16-3 and 16-4, the slopes are, respectively, .42 and 1.07.

$$r = \sqrt{(.42)(1.07)}$$
$$= \sqrt{.4494}$$
$$= .67 \qquad \text{exactly the same result}$$

## 17-12   Short Form for Computing $r$ from Grouped Data

Obviously, the methods previously described cannot be used to obtain a correlation coefficient from grouped data. It is, however, relatively easy to derive such a formula. Since

$$r = \frac{N\Sigma XY - (\Sigma X)(\Sigma Y)}{\sqrt{[N\Sigma X^2 - (\Sigma X)^2][N\Sigma Y^2 - (\Sigma Y)^2]}}$$

the corresponding computational formula for grouped data must be similar in form, merely taking into account the use of frequencies and classes. An additional refinement, however, can be included. The correlation coefficient is not affected by the units of either $X$ or $Y$. Therefore, since step-deviation units save considerable effort, they can be used without applying a correction factor for the change in units. Since two new variables are used, $d'_x$ and $d'_y$, these new symbols are substituted wherever $X$ and $Y$ occurred in the previous formula. The corresponding frequencies in classes of the $X$ variable are denoted by $f_x$, and the frequencies of $Y$ are expressed as $f_y$. The formula above becomes

$$r = \frac{N\Sigma f(d'_x)(d'_y) - (\Sigma f_x d'_x)(\Sigma f_y d'_y)}{\sqrt{[N\Sigma f_x(d'_x)^2 - (\Sigma f_x d'_x)^2][N\Sigma f_y(d'_y)^2 - (\Sigma f_y d'_y)^2]}}$$

The formula for grouped data may seem forbidding, but actual computation is relatively simple and fast. Grouped data are analyzed in Fig. 17-10 showing the relationship between scores on a test of verbal ability and ability to comprehend scientific materials. With 231 cases, the computation of the correlation coefficient from ungrouped data would require several hours of labor and the probability of making errors would be greatly increased.

Figure 17-6 shows the technique of constructing a scatter diagram or frequency table for the two variables. Since each person was scored on both variables, the tally marks represent simultaneously the scores on the two tests. The next figure (Fig. 17-7) shows the regression lines of the variables, each dot separately plotted to indicate the scatter of the scores about the regression lines. Note that the basic positive correlation between

TEST I: GENERAL VERBAL ABILITY

*TEST II: ABILITY TO COMPREHEND SCIENTIFIC MATERIAL*

| Y \ X | 10–19 | 20–29 | 30–39 | 40–49 | 50–59 | 60–69 | 70–79 | 80–89 | f_Y |
|---|---|---|---|---|---|---|---|---|---|
| 90–99 | | | | | | | I | II | 3 |
| 80–89 | | I | | | | IIII | I | II | 8 |
| 70–79 | | | | I | IIII | 卌 卌 | III | III | 21 |
| 60–69 | | | 卌 I | II | 卌卌 卌 IIII | 卌卌 卌 IIII | 卌 卌卌 IIII | II | 62 |
| 50–59 | I | II | 卌 I | 卌卌 卌卌 卌 I | 卌卌 卌卌 卌 | 卌 卌 II | 卌 | I | 73 |
| 40–49 | | IIII | 卌 卌卌 卌 | 卌 卌 I | 卌 卌 III | III | I | | 51 |
| 30–39 | | II | 卌 | I | | | | | 8 |
| 20–29 | | | III | I | | | | | 4 |
| 10–19 | | | | I | | | | | 1 |
| f_x | 1 | 9 | 40 | 43 | 55 | 43 | 30 | 10 | 231 |

Fig. 17-6 Construction of a scatter diagram. Note especially that the class intervals for the Y axis run from bottom to top, and for the X axis from left to right. The number of cases in each cell is indicated by tally marks—the frequencies for the Y variable by a special column on the right and the frequencies for the X variable by a special row at the bottom. The data represent scores on two tests (general verbal ability and comprehension of scientific material) administered to a sample of 231 freshmen in the College of Engineering, University of Washington.

the two variables is shown by either type of scatter diagram. The frequency table (Fig. 17-6) is used for computation because it provides data for later correlation analysis. Figure 17-8 shows in three-dimensional perspective the correlation surface produced by the two variables. A smoothed form of this correlation surface is illustrated in Fig. 17-9.

The computation of the correlation coefficient from grouped data is shown in Fig. 17-10. The computation can be illustrated by observing the row labeled 80–89. The classes are all of equal width, so that the row 80–89 is three step-deviation units above the mid-point of the row chosen as the guessed mean. There are eight cases in this class, so that $f_Y d_Y'$ is 24, the product of the two numbers. To obtain $f_Y(d_Y')^2$, we merely multiply the previous

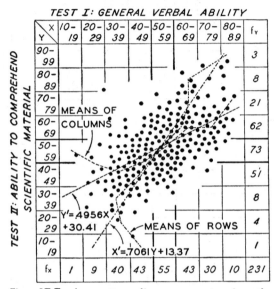

Fig. 17-7 A scatter diagram portraying the relationship between scores on a test of general verbal ability and on a test of ability to comprehend scientific material.

answer by $d_Y'$, which is three times 24. The result is 72. Summation gives the totals required in the formula.

It is slightly more difficult to compute $f_Y(d_X')(d_Y')$. This is the covariation in terms of the new variables. Each cell in the correlation table is the result of classifying the cases by both variables, so that each cell is the product of both step deviations. The second column in the 80–89 row indicates that one person within that class also has a score in general verbal ability of 20 to 29. The product of the two step deviations, which is $-9$, is written in the upper left-hand corner of the cell. The product of $-9$ and

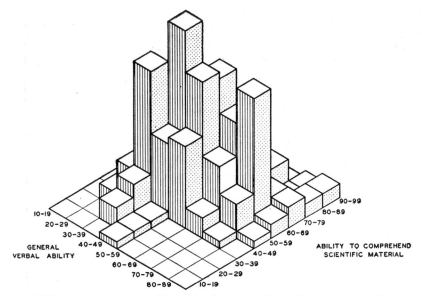

Fig. 17-8  Block diagram of a correlation surface showing the relationship between general verbal ability and ability to comprehend scientific material for a sample of 231 freshmen in the College of Engineering, University of Washington.

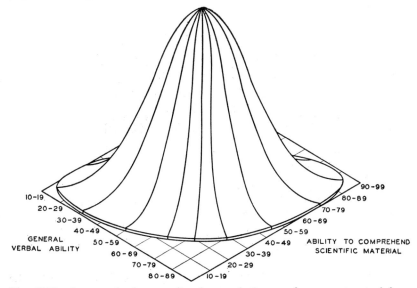

Fig. 17-9  A smoothed normalized correlation surface constructed from Fig. 17-8.

TEST I: GENERAL VERBAL ABILITY

TEST II: ABILITY TO COMPREHEND SCIENTIFIC MATERIAL

| Y \ X | 10–19 | 20–29 | 30–39 | 40–49 | 50–59 | 60–69 | 70–79 | 80–89 | $d'_Y$ | $f_Y$ | $f_Y d'_Y$ | $f_Y(d'_Y)^2$ | $f_Y(d'_X)(d'_Y)$ |
|---|---|---|---|---|---|---|---|---|---|---|---|---|---|
| 90–99 | $^{-16}$ | $^{-12}$ | $^{-8}$ | $^{-4}$ | $^{0}$ | $^{4}$ | $^{8}$ 1 | $^{12}$ 2 | 4 | 3 | 12 | 48 | 32 |
| 80–89 | $^{-12}$ | $^{-9}$ 1 | $^{-6}$ | $^{-3}$ | $^{0}$ | $^{3}$ 4 | $^{6}$ 1 | $^{9}$ 2 | 3 | 8 | 24 | 72 | 27 |
| 70–79 | $^{-8}$ | $^{-6}$ | $^{-4}$ | $^{-2}$ 1 | $^{0}$ 4 | $^{2}$ 10 | $^{4}$ 3 | $^{6}$ 3 | 2 | 21 | 42 | 84 | 48 |
| 60–69 | $^{-4}$ | $^{-3}$ | $^{-2}$ 6 | $^{-1}$ 2 | $^{0}$ 19 | $^{1}$ 14 | $^{2}$ 19 | $^{3}$ 2 | 1 | 62 | 62 | 62 | 44 |
| 50–59 | $^{0}$ 1 | $^{0}$ 2 | $^{0}$ 6 | $^{0}$ 26 | $^{0}$ 20 | $^{0}$ 12 | $^{0}$ 5 | $^{0}$ 1 | 0 | 73 | 0 | 0 | 0 |
| 40–49 | $^{4}$ | $^{3}$ 4 | $^{2}$ 20 | $^{1}$ 11 | $^{0}$ 12 | $^{-1}$ 3 | $^{-2}$ 1 | $^{-3}$ | −1 | 51 | −51 | 51 | 58 |
| 30–39 | $^{8}$ | $^{6}$ 2 | $^{4}$ 5 | $^{2}$ 1 | $^{0}$ | $^{-2}$ | $^{-4}$ | $^{-6}$ | −2 | 8 | −16 | 32 | 34 |
| 20–29 | $^{12}$ | $^{9}$ | $^{6}$ 3 | $^{3}$ 1 | $^{0}$ | $^{-3}$ | $^{-0}$ | $^{-9}$ | −3 | 4 | −12 | 36 | 21 |
| 10–19 | $^{16}$ | $^{12}$ | $^{8}$ | $^{4}$ 1 | $^{0}$ | $^{-4}$ | $^{-8}$ | $^{-12}$ | −4 | 1 | −4 | 16 | 4 |
| $d'_X$ | −4 | −3 | −2 | −1 | 0 | 1 | 2 | 3 | | 231 | 57 | 401 | 268 |
| $f_X$ | 1 | 9 | 40 | 43 | 55 | 43 | 30 | 10 | 231 | | | | |
| $f_X d'_X$ | −4 | −27 | −80 | −43 | 0 | 43 | 60 | 30 | −21 | | | | |
| $f_X(d'_X)^2$ | 16 | 81 | 160 | 43 | 0 | 43 | 120 | 90 | 553 | | | | |
| $f_X(d'_X)(d'_Y)$ | 0 | 15 | 66 | 16 | 0 | 43 | 62 | 66 | 268 | | | | |

$$r = \frac{N\,\Sigma f(d'_x)(d'_y) - (\Sigma f_x d'_x)(\Sigma f_y d'_y)}{\sqrt{[N\,\Sigma f_x(d'_x)^2 - (\Sigma f_x d'_x)^2][N\,\Sigma f_y(d'_y)^2 - (\Sigma f_y d'_y)^2]}}$$

$$= \frac{231(268) - (-21)(57)}{\sqrt{[231(553) - (-21)^2][231(401) - (57)^2]}}$$

$$= \frac{61{,}908 - (-1{,}197)}{\sqrt{[(127{,}743) - (441)][(92{,}631) - (3{,}249)]}}$$

$$= \frac{63{,}105}{\sqrt{[127{,}302][89{,}382]}}$$

$$= \frac{63{,}105}{\sqrt{11{,}378{,}507{,}364}}$$

$$= \frac{63{,}105}{106{,}670.0865}$$

$$= .5916$$

$$= .59$$

**Fig. 17-10** Correlation table illustrating computational procedure for the product-moment coefficient of correlation.

1 is $-9$. Summing for the entire row, we get the values $-9, 12, 6$, and 18. The sum is 27. Summation for all the rows gives 268 as the covariation. Using the same technique for the columns provides a check, and once again the total is 268. The check is provided because the covariation is the most likely point in the computation for the introduction of errors.

## 17-13   Rank Correlation Coefficient

A special type of correlation coefficient is based on the rankings of two variables. It is known as the *coefficient of rank correlation* and is symbolized by $\rho$ (Greek rho). The rank-order correlation technique is based on the same reasoning as the product-moment correlation coefficient. Ranks are substituted for scores used in ordinary straight-line correlation. Sometimes in social research it may be possible to rank a characteristic but not to give it a more specific value. For example, we may know that Jim is better than James, and that James is better than Jimmy, and yet be unable to give each a definite score. An alternative is to rank Jim first, James second, and Jimmy third. Although the analysis of ranks is not as precise as the analysis of scores or other original values of a variable, it is often necessary to use them. In addition, rank correlation has a further advantage which will be discussed in Sec. 17-19.

The procedure for computing a rank coefficient of correlation $(\rho)$ is relatively simple. The first step is to rank from highest to lowest the cases or items in each of the two variables. Ranking merely involves the numbering of the values in each series according to the positions they occupy when arranged in order of magnitude. The highest value is given the rank of 1, the next highest 2, and so on. The next step is to find the differences in the rankings for each of the cases. The differences in the rankings are squared and summed. The appropriate substitutions are then made in the following formula:

$$\rho = 1 - \frac{6\Sigma d^2}{N(N^2 - 1)}$$

**Illustration.** Find the rank correlation coefficient between major crimes and mobility of population in 14 states in 1950.

### Table 17-1    Calculation of Rank Correlation Coefficient

| (1) State | (2) Rank in crime rate | (3) Rank in mobility | (4) Deviations in rank between crime and mobility (d) | (5) Deviations squared (d²) |
|---|---|---|---|---|
| A | 1 | 4 | 3 | 9 |
| B | 2 | 1 | 1 | 1 |
| C | 3 | 2 | 1 | 1 |
| D | 4 | 6 | 2 | 4 |
| E | 5 | 10 | 5 | 25 |
| F | 6 | 5 | 1 | 1 |
| G | 7 | 7 | 0 | 0 |
| H | 8 | 3 | 5 | 25 |
| I | 9 | 12 | 3 | 9 |
| J | 10 | 14 | 4 | 16 |
| K | 11 | 9 | 2 | 4 |
| L | 12 | 11 | 1 | 1 |
| M | 13 | 13 | 0 | 0 |
| N | 14 | 8 | 6 | 36 |
|   |   |   |   | $\Sigma d^2 = 132$ |

$$\rho = 1 - \frac{6\Sigma d^2}{N(N^2 - 1)}$$

$$= 1 - \frac{6(132)}{14(14^2 - 1)}$$

$$= 1 - \frac{792}{14(195)}$$

$$= 1 - \frac{792}{2,730}$$

$$= .71$$

## 17-14   Tied Ranks

A problem that may arise in the computation of the rank correlation coefficient is the assignment of ranks in the case of ties. When ties occur, the tied cases are given the same ranking which represents the mean rank of the tied cases. If John and Joan tie for first place, they really occupy first and second places. If it were possible to discriminate between them, they would occupy the first two ranks. Therefore, they are both given the rank 1.5, the mean of the ranks for which they are tied. The person following them would be given the rank of 3.

Table 17-2  Ranking Procedure for Cases That Are Tied

| (1) | (2) | (3) |
|---|---|---|
| Case designation | School years completed | Rank |
| A | 3 | 1 |
| B | 4 | 2.5 |
| C | 4 | 2.5 |
| D | 5 | 5 |
| E | 5 | 5 |
| F | 5 | 5 |
| G | 7 | 7 |
| H | 8 | 8 |
| I | 9 | 9.5 |
| J | 9 | 9.5 |
| K | 10 | 11.5 |
| L | 10 | 11.5 |
| M | 12 | 13 |
| N | 16 | 14 |

The formula given above for rank correlation is most efficient when there are no tied ranks. If the proportion of tied ranks is large, say above 20 per cent, more complex formulas must be used. The formula indicated in Sec. 17-13 works well if the number of ties is not very large.

## 17-15  Direction of Ranking

The assignment of ranks can begin at the lowest number or at the highest number of a distribution. Since this choice is arbitrary, the sign of the rank correlation coefficient depends on the direction of ranking. Generally, it is advisable to follow a consistent practice, starting either at the bottom or at the top for both variables. If this practice is followed, the sign of the rank correlation coefficient will have the same meaning as the sign of the straight-line correlation coefficient.

## 17-16    Sampling Distribution of $r$

The sampling distribution of $r$ is not normal when the correlation in the universe is large or the number of cases is small. If we wish to test the null hypothesis that the correlation in the universe is 0, the sampling distributions have been studied, and the lowest significant $r$ is known. The number of degrees of freedom is $N - 2$, since there are two constants fixed in finding the least-squares line. Table D shows the lowest $r$ that indicates a significant correlation in the universe for a specified number of degrees of freedom.

**Illustration.** It was found that the correlation between alcohol consumption and income is $-.14$ in a sample of 29 workers. Does this indicate a significant relationship between drinking and income?

Table D shows that the lowest significant $r$ at 27 degrees of freedom $(29 - 2)$ is .367. This observed correlation is only $-.14$, smaller than .367. Therefore, we cannot reject the null hypothesis that there is no relationship between alcohol consumption and income. This correlation might arise through sampling variability.

## 17-17    Assumptions in Using Sampling Distribution of $r$

All tests of significance of $r$, such as illustrated above or described below, are based on three assumptions:

1. There is a straight-line relationship between the two variables (compare Fig. 17-4).

2. The frequencies should approximate a bivariate normal distribution, with most of the cases having middle values in both variables. Minor deviations from normality are not important, but extremes of skewness or kurtosis in either variable prevent the test of hypotheses concerning the correlation in the universe.

3. The variability does not change markedly as the values of either variable increase or decrease. For example, the workers with very low incomes should have about the same variability in alcoholic consumption as workers with slightly higher incomes or workers with very high incomes. The standard deviation of the classes in grouped data should not differ very greatly from class to class. This is known as *homoscedasticity*, similarity of scatter throughout the range of a variable.

Usually an inspection of the scattergram or correlation surface will indicate whether there is enough divergence from these three assumptions to give unreliable results. The correlation surface produced by the tests of verbal ability and ability to understand scientific material (Fig. 17-8) was close to a bivariate normal distribution. In contrast, the accompanying correlation surface (Fig. 17-11) shows the negative relationship between percentage of votes for Roosevelt in 1940 and socioeconomic status of 505 precincts in Seattle. The surface is markedly skewed. Skewed distributions are frequently found in social-science data, seldom conforming to the assumptions of product-moment correlation. Many researchers ignore the failure of their data to meet the assumptions of product-moment correlation, since they are not generalizing to any known population. But

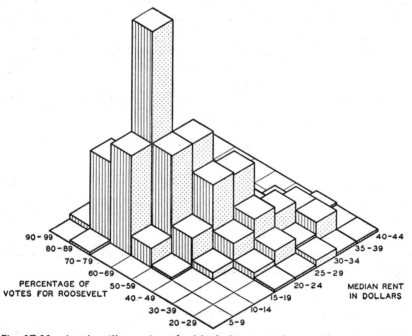

**Fig. 17-11** Another illustration of a block diagram of a correlation surface. Note the negative correlation ($-.71$) between voting strength of Roosevelt and socioeconomic status. The data represent ecological distributions based on 505 precincts. The marked skewness indicated by the distributions on this chart is not at all unusual for sociological data.

failure to fulfill these assumptions prevents testing of many theoretical propositions by the technique of product-moment correlation. In such cases, the correlation derived from the sample is true for that sample, but it cannot be used as the basis for an estimate of correlation in the universe. To meet this difficulty, various methods of rank correlation are being used increasingly in social research.

## 17-18    Fisher's $Z$

Table D presents information about the sampling distribution of $r$ when the correlation in the universe is zero. This table, however, gives no basis for construction of confidence limits for *correlation in the universe*, $r_u$. We said that the distribution of $r$ is not normal when $r_u$ is large, so that the normal curve cannot be used for tests of any hypotheses other than $r_u$ is a small value, such as 0. We need a sampling distribution that is normal regardless of the value of $r_u$. This is accomplished by changing $r$ into another variable whose sampling distribution is approximately normal for any value of $r_u$.

Table E shows the values of *Fisher's Z* which correspond to any given $r$. The formula for $Z$ is complex,

$$Z = \frac{1}{2} \log_e \frac{1 + r}{1 - r}$$

Fortunately, we can merely look in Table E and transform any $r$ into its corresponding $Z$. Then we can work with $Z$, with its normal distribution, and transform the final result back from $Z$ to $r$.

The sampling distribution of $Z$ is normal, with mean 0. The *standard error of Z* $(\sigma_z)$ is equal to

$$\frac{1}{\sqrt{N - 3}}$$

**Illustration.** In a sample of 48 cases, the correlation between years of participation in the Boy Scouts and education of the father is .37. What are the 95 per cent confidence limits for the correlation in the universe?

$$r = .37$$

The corresponding $Z$, from Table E, is .39.

$$\sigma_z = \frac{1}{\sqrt{N-3}}$$
$$= \frac{1}{\sqrt{45}}$$
$$= \frac{1}{6.71}$$
$$= .15$$
$$1.96(\sigma_z) = 1.96(.15) = .29$$

The lowest $Z$ is $.39 - .29 = .10$.
The highest $Z$ is $.39 + .29 = .68$.
In terms of $Z$ scores, the confidence limits are .10 and .68. We must now translate these $Z$ scores back into correlation coefficients.

Referring again to Table E, a $Z$ of .10 corresponds to an $r$ of .10. A $Z$ of .68 corresponds to an $r$ of .59.

The 95 per cent confidence limits for the correlation between years of participation in the Boy Scouts and education of the father are .10 and .59. The correlation in the universe is believed to lie between these limits.

## 17-19 Rank Correlation Coefficient as a Nonparametric Measure

In problems where the assumptions of Sec. 17-17 cannot be matched by the data, it is often advisable to use coefficients of rank correlation. The rank correlation coefficient makes no assumptions about the universe from which the sample is taken. It is therefore a *nonparametric measure*, since it makes no assumptions about the parameters in the universe. Such measures are also called *distribution-free*.

## 17-20 Sampling Distribution of $\rho$

When $N$ is greater than 20, the null hypothesis that the rank correlation coefficient in the universe is 0 can be tested. The sampling distribution when the population rank correlation coefficient is 0 forms a normal curve with a mean of 0 and a *standard error of the coefficient of rank correlation* ($\sigma_\rho$) equal to

$1/\sqrt{N-1}$. Except for the test that the rank correlation coefficient in the population is 0, no tests of the significance of $\rho$ are possible.

**Illustration.** The rank correlation coefficient between degree of segregation and proportion of population classified as Negro in 30 states is .44. Is there a significant association between segregation and the proportion of the population that is Negro?

$$\sigma_\rho = \frac{1}{\sqrt{N-1}}$$
$$= \frac{1}{\sqrt{29}}$$
$$= .186$$
$$z = \frac{.44 - 0}{.186}$$
$$= 2.37$$

Since 2.37 is greater than 1.96, this sample demonstrates that there is a positive correlation in the universe between segregation and proportion of the population that is Negro.

## 17-21    Correlation and Causation

Even if statistically significant, a correlation coefficient must not be interpreted as proof of causation. First, the correlation coefficient is a measure of covariation. It does not indicate which variable is the cause and which is the effect. If among scientists there is a negative correlation between his reputation and the age at which the scientist published his first research paper, there is no clear guide as to which is the causative variable. Were the men who became famous so superior in ability that their earliest work was deemed worthy of being published, or was the early publication an aid to them in becoming well-known? The correlation coefficient cannot answer this question.

Second, other uncontrolled factors affecting the variables may be the true causative factors. If there is a positive correlation between age and the ability to avoid being involved in automobile accidents, there are many additional factors which may explain the correlation. For example, younger drivers do more dating, thereby exposing themselves to the greater risk of driving at night. Another explanation is that younger drivers may have less awareness of financial responsibility in case of an accident. Perhaps the basic variable is the lack of experience of the younger drivers. These and many other variables could explain the correlation of driving ability with age. In this, as in other social research, techniques are no substitute for thinking. The researcher who has intimate understanding of his subject matter is more likely to choose crucial variables for his investigation, rather than grinding out meaningless correlations.

## 17-22  Summary

The covariation is influenced by the sign and size of the relationship between two variables.

The coefficient of correlation ranges from $-1$ to $1$.

$$r^2 = \frac{\text{explained variation}}{\text{total variation}} = 1 - \frac{\text{unexplained variation}}{\text{total variation}}.$$

$$r = \sqrt{b_{xr}b_{yx}}.$$

$$\rho = 1 - \frac{6\Sigma d^2}{N(N^2 - 1)}.$$

In deriving the sampling distribution of $r$, the following assumptions are made:

1. Straight-line relationship
2. Approximately bivariate normal distribution
3. Homoscedasticity

The rank correlation coefficient is a nonparametric measure.

The sampling distribution of $r$ is not normal when $r_u$ is large or $N$ is small.

The $Z$ transformation produces a sampling distribution which is relatively independent of $r_u$. It is normal in form, with mean $0$ and

$$\sigma_z = \frac{1}{\sqrt{N - 3}}$$

When testing the hypothesis that the rank correlation coefficient in the universe is $0$,

$$\sigma_\rho = \frac{1}{\sqrt{N - 1}}$$

## PROBLEMS FOR CHAP. 17

**1.** The unexplained variation between grades and extracurricular activities is 41, out of a total variation of 112. What is $r$?   *Ans.* .80

**2.** The slope of the least-squares line is .65 when cohesiveness of the group is the independent variable and rejection of out groups is the dependent variable. When cohesiveness is viewed as the dependent variable, $b = .44$. What is the correlation between the two variables? If there are 42 members in the group, is the correlation significantly above zero?                               *Ans.* .53 is significant

**3.** Is a rank correlation coefficient of .25 significantly above zero when it is based on a sample of 100 cases?                 *Ans.* Yes

**4.** Given the correlation table below, compute the correlation between accurate perception of others and accurate perception of self.
                                                                    *Ans.* .27

PERCEPTION OF SELF

| X / Y | 0 – 1 | 2 – 3 | 4 – 5 | 6 – 7 | 8 – 9 | 10 – 11 | 12 – 13 | 14 – 15 | f |
|---|---|---|---|---|---|---|---|---|---|
| 9 |  |  |  |  |  |  | 1 | 1 | 2 |
| 8 | 1 |  | 2 | 1 |  | 2 | 2 | 3 | 11 |
| 7 | 2 | 1 | 3 | 5 | 9 | 3 | 2 | 4 | 29 |
| 6 | 5 | 5 | 3 | 6 | 8 | 3 | 4 | 5 | 39 |
| 5 | 4 | 5 | 1 | 11 | 10 | 2 | 2 | 4 | 39 |
| 4 |  | 5 | 4 | 5 | 6 | 1 |  | 3 | 24 |
| 3 |  | 4 | 4 | 2 | 4 |  |  |  | 14 |
| 2 |  | 1 | 1 |  |  |  |  |  | 2 |
| f | 12 | 21 | 18 | 30 | 37 | 11 | 11 | 20 | 160 |

PERCEPTION OF OTHERS

# 18 Contingency

VOCABULARY

contingency
independence values
marginal totals
2 × 2 table

chi square ($\chi^2$)
phi coefficient ($\phi$)
coefficient of contingency ($C$)
Tschuprow's $T$ ($T$)
coefficient of relative predictability ($G$)

The preceding chapter was concerned with correlation between two quantitative variables. This chapter will be devoted to a discussion of correlation between two variables in qualitative form. Since several techniques have been devised for measuring the correlation of qualitative variables, it is essential to know the strength and weakness of each technique.

## 18-1 Contingency and Qualitative Data

As was indicated in a previous discussion (Sec. 7-1), there is no sharp division between quantitative and qualitative variables. Quantitative data can be made qualitative, and qualitative data can be changed into quantitative form. In view of these facts it is reasonable to expect that methods would have been developed to measure the correlation between two qualitative variables, just as the correlation coefficient has been devised to measure relationships between quantitative data. Contingency refers to the idea of correlation applied to nonquantitative data. The principles of contingency developed here are analogous to, but not identical with, the basic principles of quantitative correlation.

Table 18-1 shows the relation of color to type of employment in a small Southern community. According to these data white males are more likely than nonwhite males to have white-collar occupations, while nonwhite males are more likely to have blue-collar jobs. There is a relationship between color and type of job, but as yet we have no measure to express the strength of the relationship.

## 18-2  Independence Values

Let us suppose that there is no connection between color and type of job. On the basis of this supposition, we could compute a theoretical number of whites and nonwhites in each type of job. These would be the *independence values*, the frequencies that would be found if there were no relationship between the two variables. The greater the divergence of observed frequencies from theoretical frequencies, the stronger is the relationship between two variables.

In order to compute the independence values, one must know the *marginal totals*, the total number of cases for each classification. The marginal totals in Table 18-1 are 250, 150, 100, and 300. From the marginal totals, it is clear that one out of every four workers had a white-collar job (100 out of 400). If there were no relation between color and type of job, then $\frac{1}{4}$ of the white males should have white-collar jobs. One-quarter of 250 is 62.5, the independence frequency of white males with white-collar jobs. Correspondingly, 37.5 of the nonwhites should be white-collar workers, since $\frac{1}{4}$ of 150 is 37.5. If $\frac{1}{4}$ of all workers had white-collar jobs, then obviously $\frac{3}{4}$ of them should have

Table 18-1   Relationship between Color and Type of Job for Male Workers in a Small Southern Community

| Type of job | White males | Nonwhite males | Total, all males |
|---|---|---|---|
| White-collar | 90 | 10 | 100 |
| Blue-collar | 160 | 140 | 300 |
| Total, all jobs | 250 | 150 | 400 |

Table 18-2  Independence Values and Observed Frequencies Showing Relationship between Color and Type of Job for Male Workers in a Small Southern Community

| Type of job | White males | Nonwhite males | Total, all males |
|---|---|---|---|
| White-collar | (62.5) 90 | (37.5) 10 | 100 |
| Blue-collar | (187.5) 160 | (112.5) 140 | 300 |
| Total, all jobs | 250 | 150 | 400 |

blue-collar jobs. The independence value for whites in blue-collar jobs would be 187.5 ($\frac{3}{4}$ of 250), and for nonwhites in such jobs it would be 112.5 ($\frac{3}{4}$ of 150).

A more direct method of finding the independence values takes advantage of our knowledge of the marginal totals. In a $2 \times 2$ *table*, one which has two classifications for the columns and two classifications for the rows, the computation of only one independence value by the method outlined above would enable us to know immediately the other independence values. Table 18-2 includes the data of Table 18-1 in addition to the independence values which are in parentheses. Knowing any one of the independence values, we can find the others by subtraction. For example, the independence value for white males in white-collar positions is 62.5. With this information, let us compute the other values. There are 250 white males, so that 187.5 (250 minus 62.5) must be white blue-collar workers. There are 100 white-collar workers, so 37.5 (100 minus 62.5) must be nonwhite white-collar workers. Of the 150 nonwhite males, 112.5 (150 minus 37.5) must be blue-collar workers.

## 18-3   Chi Square

The greater the difference between the theoretical independence frequencies and the observed frequencies, the stronger the relationship between the two variables: color and type of job. We now need a measure of the discrepancy between the observed and independence frequencies. The usual measure of such divergence is $\chi^2$ (*chi square*). It is defined by the following formula:

$$\chi^2 = \Sigma \frac{(\text{observed frequency} - \text{theoretical frequency})^2}{\text{theoretical frequency}}$$

$$= \Sigma \frac{(f_0 - f_t)^2}{f_t}$$

A simple table (Table 18-3) will suffice for the computation of chi square.

Table 18-3    Computation Table for Finding $\chi^2$ from Table 18-2

| $f_0$ | $f_t$ | $f_0 - f_t$ | $(f_0 - f_t)^2$ | $\dfrac{(f_0 - f_t)^2}{f_t}$ |
|---|---|---|---|---|
| 90 | 62.5 | 27.5 | 756.25 | 12.10 |
| 160 | 187.5 | −27.5 | 756.25 | 4.03 |
| 10 | 37.5 | −27.5 | 756.25 | 20.17 |
| 140 | 112.5 | 27.5 | 756.25 | 6.72 |

$$\chi^2 = 43.02$$

The larger the chi square, the stronger the relationship between the two variables. Unfortunately, this measure is affected by the number of cases being analyzed. A large $N$ is likely to produce a large $\chi^2$. Various techniques have been devised to meet this problem, including attempts to find some pure number that will express the degree of relationship. Each of these measures is based on chi square, a very useful statistical tool. The only prerequisites for the use of chi square are that the two variables be continuous (Sec. 2-4) and that no theoretical frequency be less than 5. (Some statisticians demand a minimum theoretical frequency of 10.) The assumption of continuity is often violated without serious results. Correction factors are available to adjust for lack of continuity, but the usefulness of such corrections is a matter of dispute. It should be pointed out, however, that chi square must not be used when there is any theoretical frequency below 5.

## 18-4    Phi Coefficient

The first measure of contingency is $\phi$ (*phi*). It is defined by the following formula:

$$\phi = \sqrt{\frac{\chi^2}{N}}$$

It is usually used to measure relationship for $2 \times 2$ tables, where each variable is split into two parts or into two distinct classes. Applied to our data on job types and color,

$$\phi = \sqrt{\frac{43.02}{400}} = \sqrt{.1076} = .33$$

The phi coefficient can range in size from 0 to 1. A $\phi$ of zero would occur only when chi square is 0, indicating that there is no difference between observed frequencies and independence values. A $\phi$ of 1 would represent the strongest possible relationship between two variables. The direction of relationship is not indicated and must be noted by examination of the observed and theoretical frequencies.

## 18-5    Coefficient of Contingency

Another common measure of the strength of relationship is the *coefficient of contingency*, $C$. It is defined by

$$C = \sqrt{\frac{\chi^2}{N + \chi^2}}$$

The lower limit of $C$ is 0, but the upper limit shifts according to the number of categories in the two variables. In a $2 \times 2$ table, the upper limit is .707, while in a $5 \times 5$ table the limit becomes .894. A $10 \times 10$ table would have an upper limit of .949. Clearly, coefficients of contingency computed from different systems of categories are therefore not directly comparable. Accordingly, the coefficient of contingency should seldom be used unless the data can be divided into at least 5 categories for each variable.

## 18-6    Relation of $\phi$ and $C$

Both $\phi$ and $C$ are defined in terms of $\chi^2$ and $N$, so that there must be a direct relationship between them. In a $2 \times 2$ table,

$$C = \sqrt{\frac{\phi^2}{1 + \phi^2}}$$

**Proof.** Prove that $C = \sqrt{\phi^2/(1 + \phi^2)}$ in a $2 \times 2$ table.

$$C = \sqrt{\frac{\chi^2}{N + \chi^2}}$$

Let us divide both the numerator and the denominator by $N$.

$$C = \sqrt{\frac{\chi^2/N}{(N + \chi^2)/N}}$$

In the denominator we have $N$ divided by $N$, which is 1. Therefore,

$$C = \sqrt{\frac{\chi^2/N}{1 + \chi^2/N}}$$

Now $\phi$ is equal to

$$\sqrt{\chi^2/N}$$

so that $\phi^2$ must be the same as $\chi^2/N$.
Wherever $\chi^2/N$ appears in the last formula for $C$, we can substitute $\phi^2$. Therefore,

$$C = \sqrt{\frac{\phi^2}{1 + \phi^2}} \qquad \text{for a } 2 \times 2 \text{ table.}$$

## 18-7    Tschuprow's $T$

*Tschuprow's $T$* is a measure of contingency which attempts to eliminate the major deficiency of $C$, its dependence on the number of categories. It is completely successful only where the number of columns equals the number of rows. It is defined according to the following formula:

$$T^2 = \frac{\chi^2}{N\sqrt{(s - 1)(t - 1)}}$$

where $s$ is the number of categories for one variable
$t$ is the number of categories for the other

## 18-8   Degrees of Freedom

It is obvious that Tschuprow's $T$ seeks to take into account the number of categories involved in the cross classification. But the reason for subtracting 1 from both the number of rows and the number of columns is not so apparent. It involves the concept of degrees of freedom, the number of observations that are free to vary (see another application in Sec. 15-3). In Sec. 18-2 we noted that in a 2 × 2 table, if we know one observed or theoretical frequency, we can derive all the other frequencies by subtracting from the marginal total. Since the number of categories for each variable is 2, the product of $(s-1)$ and $(t-1)$ would be 1, the correct number of degrees of freedom. We can conclude that all 2 × 2 tables in which we know the marginal totals have 1 degree of freedom, since the other frequencies are fixed by the value of any single frequency.

In a 3 × 4 table, the number of degrees of freedom is

$$(s-1)(t-1) = (3-1)(4-1) = (2)(3) = 6$$

To prove this result, let us designate by $X$ the six cells of the 3 × 4 table that determine the frequencies of the remaining cells. Then we shall see the way the other frequencies are fixed by the values found in the six cells that are free to vary.

| | | | | |
|---|---|---|---|---|
| $X$ | $X$ | $X$ | $a$ | Marginal total |
| $X$ | $X$ | $X$ | $b$ | Marginal total |
| $c$ | $d$ | $e$ | $f$ | Marginal total |
| M.T.* | M.T. | M.T. | M.T. | Grand total |

\* Marginal total

The frequency in cell $a$ can be found by subtracting the sum of the other cells in the first row from the marginal total. The frequency in cell $b$ is fixed in the same way, since, whatever values are found for the other cells in the row, the number in $b$ must be just exactly the amount necessary to produce the marginal total.

**Table 18-4** Relationship between In-law Relations and Marital Happiness for 250 Couples

| Marital happiness | In-law relations | | | |
|---|---|---|---|---|
| | Excellent | Good | Fair or poor | Total |
| Low | (20) 8 | (16) 12 | (14) 30 | 50 |
| Average | (40) 20 | (32) 51 | (28) 29 | 100 |
| High | (40) 72 | (32) 17 | (28) 11 | 100 |
| Total | 100 | 80 | 70 | 250 |

**Table 18-5** Computation Table for Finding $T$ from Table 18-4

| $f_0$ | $f_t$ | $f_0 - f_t$ | $(f_0 - f_t)^2$ | $\dfrac{(f_0 - f_t)^2}{f_t}$ |
|---|---|---|---|---|
| 8 | 20 | −12 | 144 | 7.20 |
| 12 | 16 | − 4 | 16 | 1.00 |
| 30 | 14 | 16 | 256 | 18.29 |
| 20 | 40 | −20 | 400 | 10.00 |
| 51 | 32 | 19 | 361 | 11.28 |
| 29 | 28 | 1 | 1 | .04 |
| 72 | 40 | 32 | 1024 | 25.60 |
| 17 | 32 | −15 | 225 | 7.03 |
| 11 | 28 | −17 | 289 | 10.32 |

$$\chi^2 = 90.76$$

$$T^2 = \frac{\chi^2}{N\sqrt{(s-1)(t-1)}} = \frac{90.76}{250\sqrt{(2)(2)}} = \frac{90.76}{500} = .1815$$

$$T = \sqrt{.1815} = .43$$

In similar fashion, the frequencies of $c$, $d$, and $e$ are found by subtracting the sum of the other two cells in each column from the marginal total for that column. The frequency of $f$ is fixed because $c + d + e + f$ must equal the marginal total for the third row, and we now know $c$, $d$, and $e$. The frequency of $f$ is also determined by the fact that $a$ and $b$ are known and that $a + b + f$ must equal the marginal total for the fourth column.

**Illustration.** Let us determine Tschuprow's $T$ for the relationship between marital happiness and relations with in-laws for a group of married couples. (See Tables 18-4 and 18-5.)

## 18-9 Chi Square and Statistical Inference

All measures of contingency are based on frequencies. Statistical inference can be applied to such data by using $\chi^2$, which is based on the deviation of an observed frequency from a theoretical frequency. If the frequency according to some hypothesis is thought of as the theoretical frequency, then $\chi^2$ directly measures deviation of samples around some theoretical population.

In order to test the significance of $\chi^2$, we need to know how often a $\chi^2$ of a specified size would occur by chance in the long run. The sampling distribution of $\chi^2$ is easily used for such purposes, since all we need to know is the number of degrees of freedom. Table F gives the values of $\chi^2$ for a given level of significance at each number of degrees of freedom. A $\chi^2$ larger than the value in the table is significant and would cause rejection of the hypothesis.

In a $2 \times 2$ table, there is 1 degree of freedom, since $(s - 1)(t - 1) = 1$. In Table F, a $\chi^2$ of 3.841 or larger will occur 5 per cent of the time as a result of sampling variability when there is 1 degree of freedom. Any $\chi^2$ for a $2 \times 2$ table that is larger than 3.841 indicates a significant difference that would cause rejection of the null hypothesis.

The independence values used in the calculation of $C$, $\phi$, and $T$ are based on the null hypothesis that there is no relation between the variables. A significant $\chi^2$ would cause the rejection of this null hypothesis, indicating that these variables are related.

Therefore, testing the significance of $\chi^2$ in terms of deviations from the independence values automatically tests the hypotheses that $\phi$ is 0, $C$ is 0, and $T$ is 0. Each of these measures of contingency is based on chi square, so that a significant $\chi^2$ indicates association between the variables in the population from which we sample.

- **Illustration.** In a $3 \times 4$ table, $\chi^2$ is found to be 11.4 when computed from the independence values. Is there a significant association between the two variables?

The degrees of freedom are

$$(s - 1)(t - 1) = (3 - 1)(4 - 1)$$
$$= 2(3)$$
$$= 6$$

At 6 degrees of freedom, $\chi^2$ must be at least 12.592 to be significant. We cannot reject the null hypothesis that these deviations arose by chance. No significant association between the variables has been demonstrated.

If we have two or more sets of sample frequencies and want to know if they could come from the same population, the samples can be combined to derive independence frequencies. Then $\chi^2$ can be used to measure deviations from these independence frequencies, and tests of significant differences can be made.

**Illustration.** In a sample of 100 men who take a test on knowledge of etiquette, 47 receive passing scores, whereas in a sample of 200 women who take the same test 163 pass. On the basis of these results can we conclude that women are better informed concerning rules of etiquette? By combining the results of both samples, it will be observed that 210 out of 300 persons taking the test received passing scores. The independence proportion passing is .70, and the independence proportion who fail is .30. For men, the independence frequencies are 70 and 30, while for women they are 140 and 60.

$$\chi^2 = {}^{529}\!/_{70} + {}^{529}\!/_{30} + {}^{529}\!/_{140} + {}^{529}\!/_{60}$$
$$= 7.56 + 17.63 + 3.78 + 8.81$$
$$= 37.78$$

With 1 degree of freedom for a $2 \times 2$ table, a $\chi^2$ this large would occur by chance less than 5 per cent of the time. Accordingly, it can be stated that women know more about etiquette than men, as measured by this test.

## 18-10    Using Chi Square to Test Normality

Since $\chi^2$ can measure deviations from any theoretical frequency distribution, it can be used to determine if a particular frequency distribution could arise by chance from a normally distributed population. By referring to the table of areas under the normal curve, we observe the frequencies that would be found in a normal curve with the same number of cases, same mean, and same standard deviation. We would not expect any sample to fit exactly the normal distribution, so that minor discrepancies between theoretical frequencies and observed frequencies would occur through sampling variability. $\chi^2$ can be used to test the significance of these deviations if classes are combined to ensure five cases or more in each cell of the theoretical distribution. The degrees of freedom are the number of classes minus 3, since 3 degrees of freedom are lost in matching the number of cases, mean, and standard deviation (Sec. 18-8).

**Illustration.** With seven class intervals and 300 cases, the deviation of the observed frequencies in a sample from the theoretical frequencies in a normal distribution produces a $\chi^2$ of 3.1. Could this sample come from a normal population?

$$\text{Degree of freedom} = 7 - 3 = 4$$

At 4 degrees of freedom, the lowest significant value of $\chi^2$ is 9.488. These deviations are not significant. We cannot reject the hypothesis that this sample comes from a normal population.

## 18-11    Coefficient of Relative Predictability

A measure which is gaining increased acceptance in social research is Guttman's *coefficient of relative predictability* (*G*). It is a measure of the relative efficiency of a predictive variable. If, for example, 40 per cent of all marriages in a sample are unsuccessful, most marriages (60 per cent) are successful. Therefore, the modal category represents success, and, lacking other information, the best prediction that could be made for each couple in this sample is success. The prediction, however, would be incorrect 40 per cent of the time, since success is predicted for every couple, but 40 per cent of the couples actually fail. The coefficient of relative predictability measures the extent to which the use of other predictive variables reduces this 40 per cent error in prediction.

Table 18-6 shows the number of divorces and nondivorces among a sample of 235 couples, subdivided according to their adjustment during the engagement period. If adjustment in the engagement period is related to future marital success, then predictions based on that predictive variable should give fewer errors than predictions based on the marginal modal category. In this illustration there are 125 nondivorced couples and 110 divorced couples, so the modal prediction based on the marginal totals is nondivorce. There are 110 errors.

Table 18-6   Adjustment during the Engagement Period and Marital Success for 235 Couples

| Adjustment in engagement | Marital success | |
|---|---|---|
| | Divorced | Nondivorced |
| Low | 30 | 10 |
| Average | 40 | 45 |
| High | 40 | 70 |
| Total | 110 | 125 |

When divorce or nondivorce is predicted on the basis of adjustment in engagement, the number of errors in prediction is reduced. Among those rated low in adjustment during engagement, 30 couples were divorced, and 10 were not divorced. The modal category is divorce, so that divorce is predicted for all members of this low-adjustment group. That produces 10 errors, since 10 of the couples in the low-adjustment group were not divorced.

For the average-adjustment group, the modal category is non-divorce, leaving 40 errors. For the high-adjustment group, prediction of the modal category of nondivorce again produces 40 errors. The total number of errors is $10 + 40 + 40 = 90$. The coefficient of relative predictability expresses the change in the number of errors as a proportion of the original number of errors when predicting from the marginal totals.

$$G = \frac{\text{original number of errors} - \text{final number of errors}}{\text{original number of errors}}$$

$$G = \frac{110 - 90}{110} = \frac{20}{110} = .18$$

Use of the coefficient of relative predictability has produced many startling results. For example, many parole prediction tables were found to have a low or negative $G$. Since this measure is becoming increasingly popular, it should be pointed out that $G$ must not be used uncritically. For example, a prediction table can produce more easily a high positive $G$ where marginal totals are almost equal and there is a relatively high original error. In addition, the use of only modal categories may conceal the existence of considerable correlation between the predictive instrument and the variable to be predicted. If the low-adjustment group in the above example had resulted in 20 divorces instead of 10, the predictive table would not have produced the slightest reduction in error. Finally, there is no known sampling distribution for $G$, so that no tests of significance are possible.

## 18-12 Contingency Compared to Correlation

Since quantitative data can always be changed to qualitative form and many social-science data are qualitative, the methods described in this chapter are of wide utility. They often provide a quick measure of correlation when it is not advisable to compute a product-moment correlation coefficient. Although none of the contingency measures can be directly compared to $r$, they do provide some indication of the degree of relationship between two variables.

## 18-13  Summary

$$\chi^2 = \sum \frac{(f_0 - f_t)^2}{f_t}.$$

Degrees of freedom in a table $= (s - 1)(t - 1)$.

The phi coefficient is used as a direct measure of contingency in a $2 \times 2$ table.

$C$ has a shifting upper limit, and should not be used for tables less than $5 \times 5$.

$T$ corrects for the number of categories in a square table, where the number of rows equals the number of columns.

$\chi^2$ and the number of degrees of freedom can be used to evaluate the significance of differences between frequencies.

The significance of $\phi$, $T$, and $C$ can be studied by determining if the $\chi^2$ for deviations from the independence values could result from sampling variability.

$\chi^2$ can also be used to measure the divergence from any theoretical distribution of frequencies, so that tests can be made of the shape of the population from which a sample is drawn.

$$G = \frac{\text{original number of errors} - \text{final number of errors}}{\text{original number of errors}}.$$

## PROBLEMS FOR CHAP. 18

1. When they were in love, a large sample of college students were asked whether being in love was a matter of great importance to them. Let us examine the responses in terms of the sex of the respondent. Is there a significant difference between the sexes?

| Sex | Total | Very important | Somewhat important | Not important |
|---|---|---|---|---|
| Total | 900 | 620 | 220 | 60 |
| Male | 500 | 300 | 160 | 40 |
| Female | 400 | 320 | 60 | 20 |

*Ans.* Yes

**2.** Among youths apprehended for delinquent behavior in a large Northern city, there are racial differences in the type of crime committed.

| Race | Total | Auto thefts | Other crimes |
|---|---|---|---|
| Total | 5,000 | 2,300 | 2,700 |
| Whites | 3,000 | 1,800 | 1,200 |
| Nonwhites | 2,000 | 500 | 1,500 |

*a.* Find chi square. Is there a significant color difference?

*Ans.* 592; significant

*b.* Calculate $\phi$.          *Ans.* .34

*c.* Calculate Tschuprow's $T$.          *Ans.* .34

# 19  A Taste of Things to Come

In this chapter an attempt will be made to discuss in general terms additional techniques and principles usually covered in more advanced courses in statistics.

## 19-1  Advanced Work in Statistics

A considerable portion of the technical material in this elementary course is not directly applicable to many research problems. For example, the discussions pertaining to sampling distributions in this text are based on simple random samples. Yet simple random samples are seldom available for social research. Nevertheless, the fundamental ideas which have been presented are found even in the most advanced statistical work. It is important at this stage for the student to assess correctly his ability to understand the basic principles of statistics without overevaluating his ability to handle practical problems in sampling and statistical inference.

In order to indicate the value of additional training in statistics, the remainder of this chapter will be devoted to a few examples of statistical tools which are developed in more advanced courses. The logic of statistics is fundamentally unchanged in these more advanced techniques, so that their basic significance can be understood without involvement in technical details. Advanced work in statistics will not only make available more intricate and refined tools, but it will also provide the wide range of knowledge necessary for the manipulation of social-research data.

## 19-2  Multiple Correlation

*Multiple correlation* determines the efficiency with which a variable can be predicted on the basis of more than one predictive factor. For example, high-school grades, IQ, and aspiration for high scholarship may each be positively correlated with grades in college. Each in turn can aid in predicting the college grades of any particular student. The correlation coefficients may be .60, .43, and .37, respectively. If each of these factors can be efficiently used for prediction, it is obvious that predictions of the criterion variable, grades in college, would be better if based simultaneously on all three factors.

Since the correlations between the three predicting variables and the criterion variable are all relatively high, it might be assumed that the multiple correlation coefficient would be very high. Such an assumption may not be warranted. If the predictive variables were uncorrelated with each other, then the sum of the proportions of the variance explained would be .36 + .18 + .14 = .68. The multiple correlation coefficient would be .82. The fact is, however, that seldom in social research are the predictive variables uncorrelated with each other. Accordingly, their correlation with each other means that there is an overlap in their predictive efficiency. The multiple correlation coefficient takes this into account, and many times the multiple correlation coefficient is not much higher than the best single predictive variable.

## 19-3  Partial Correlation

*Partial correlation* is to some extent the reverse of multiple correlation. It seeks to measure separately the relationship between two variables in such a way that the effects of other related variables are eliminated. For example, there is a high positive correlation between age and spelling ability. There are also high positive correlations between age and class in school and between class in school and spelling ability. Partial correlation enables the investigator to unscramble these relationships to determine if age is related to spelling ability when the effect of the class in school is eliminated. The result of this process is often a reversal of the sign of the original correlations. In this illustration, when grade in school is held constant, older children may be poorer spellers.

Partial correlation is a statistical procedure which is analogous to the controlled experiment. However, the logic of partial correlation requires a theoretical reason for eliminating the effect of certain variables. The technique cannot be applied blindly. If both parental education and income are positively correlated with the IQ of children, the researcher can effectively reduce either correlation to almost zero by eliminating the effect of the other variable. With the effect of education removed, there is almost no correlation between income and IQ. With the removal of the effect of income, the correlation of education and IQ becomes almost zero. It is not proper to conclude that the original high correlations are spurious. Both education and income are indices of socioeconomic status, and their overlap prevents either one from having a significant correlation with IQ when the effect of the other is eliminated. The user of partial correlation, to the same degree as all practitioners of statistical analysis, must relate his technique to the scientific theories which guide his research.

## 19-4   Guttman Scale

A major advance in the measurement of attitudes has been introduced by Louis Guttman. This basic technique, known as the *Guttman scale*, is based on the view that the responses of an individual to a series of items should be predictable from his total score. To illustrate, students in a class were asked to rate three items, with a score of 1 indicating agreement with an item, and 0 disagreement. The items were:

1. Learning a foreign language is very worthwhile.
2. Foreign languages should be taught in school.
3. It is nice to know at least a few phrases in a foreign language.

If this series of statements form a Guttman scale, the total scores received by individuals enable one to predict the responses on individual items. Obviously, persons scoring 0 agreed with none of the items, while those scoring 3 agreed with all of them. But those scoring 2 can also have their responses predicted. They agreed with two of the items and disagreed with one item. Most of the persons scoring 2 disagreed with the first item, the most extreme positive view, and agreed with the second and third items. Those who scored 1 are likely to have agreed only with the least positive item, item 3, disagreeing with the first and second items. Therefore, given the total score of an individual on these three questions, ranging from 0 to 3, the analyst can accurately predict the responses of most persons on the individual items.

The Guttman scale has direct application to many problems in creating tests and indices. For example, if the items form a Guttman scale, the researcher can be relatively sure that he is measuring only one thing, not a combination of several dimensions. If some of the items are difficult to understand or are unrelated to the other items, the responses to these questions will not be as predictable from the total scores. This will enable the analyst to discover which questions should be eliminated. In a sense, the employment of Guttman-scale techniques provides internal checks on the utility of the scale.

## 19-5    Analysis of Variance

*Analysis of variance* provides a procedure for testing the significance of the differences among a set of means, in which every combination of means is considered simultaneously. This is unlike the problem discussed in Chap. 15, since the differences between only a single pair of means could be analyzed by the techniques of that chapter. For example, if attitude toward nonconformists in a group were studied in a random sample of a school population, analysis of variance might be used to test the differences among four subgroups which were subjected to different treatments. The subgroups had a varying number of nonconformists, ranging from 1 to 4. The mean hostile attitude in each of the 4 groups was, respectively, 43, 38, 39, and 32. Analysis of variance can test the null hypothesis that the number of nonconformists in the groups had no effect on the feeling of hostility toward nonconformists. This more effectively answers the questions of the researcher than a test of the difference between groups 1 and 2, another test of the difference between groups 1 and 3, and so forth.

The logical foundations of analysis of variance have already been discussed. The standard error of the means is equal to the standard deviation in the universe divided by $\sqrt{N}$.

$$\sigma_{\bar{x}} = \frac{\sigma}{\sqrt{N}}$$

If the subgroups come from the same population, as stated in the null hypothesis, then the standard deviations of the means of the subgroups can be used as an estimate of the standard error of the means. Using the formula above, this estimate can be used to derive the estimated standard deviation of the population. All that is necessary is multiplication by the square root of $N$. The standard deviation of the population can also be estimated from the standard deviation of all the cases in the sample, combining the cases in the four subgroups.

The two estimates of the standard deviation in the universe can be directly compared. If the subgroups differ considerably from one another, then the standard deviation of the means will be large. This in turn will lead to a very high estimate of the standard deviation in the universe. The greater the ratio of the estimate based on the subgroup means to the estimate based on combining all the cases, the less likelihood that differences among subgroup means are due to random variability. Tests of significance can be applied to this type of measure.

Analysis of variance is a very sensitive technique, allowing the use of complex experimental designs and factoring out the effect of many variables. Unfortunately, its use in social research is circumscribed by the assumptions which underlie it, such as homoscedasticity. One of its assumptions, random assignment of the cases to the different subgroups, effectively limits its use in studies which merely observe an already existent process. This powerful tool as yet has found little application outside of controlled experiments.

## 19-6    Matching

*Matching* is a technique in which two or more groups are measured separately under different conditions, but the groups are alike in certain relevant ways. For example, irritability may be measured before and after a sample of adult males are placed on a strict diet. Each member of the group is measured before and after, so that there is matching of the persons in the two groups. Another type of matching is illustrated by setting up pairs of persons of the same ability in art and then placing one member of each pair in two groups using different methods of teaching art. The groups are matched in art ability. Finally, groups may be matched with respect to total measures, such as trying different methods of leadership on groups which have the same mean age as well as standard deviation of age.

To the extent that matching is relevant to the object of study, its effect is to make the matched groups more alike. By means of this technique the investigator can control certain variables so that their effects can be eliminated. For example, when attempting to analyze the results of differential treatment of groups, the degree of similarity of the groups must be taken into account. Since the groups are matched, differences between them are more likely to be significant, not the result of chance deviations. In more advanced work, the student will learn how to adjust his formulas based on random sampling in order to allow for the effect of matching.

The social sciences are increasingly using matched samples in statistical analysis. As the body of research findings grows, important discoveries are being made on the effects of various types of matching, the relation of matched groups to the logic of probability sampling, and the analytical difficulties produced by incomplete matching. Cautious, careful analysis is, of course, a necessary step in the execution of research using matched groups.

## 19-7    Suggestions for Further Reading

The reading of this primer will make it easier to dip into the vast literature on statistics and research methods. At this point, therefore, a few selected works are listed as possible next steps in a continuing reading program. There are many excellent books in the field, so that this brief catalogue makes no pretense to be exhaustive.

*General research methodology:* (1) Leon Festinger and Daniel Katz, *Research Methods in the Behavioral Sciences* (New York: The Dryden Press, Inc., 1953). (2) William J. Goode and Paul K. Hatt, *Methods in Social Research* (New York: McGraw-Hill Book Company, Inc., 1952). (3) Marie Jahoda, Morton Deutsch, and Stuart W. Cook, *Research Methods in Social Relations* (New York: The Dryden Press, Inc., 1951). (4) Pauline V. Young, *Scientific Social Surveys and Research* (New York: Prentice-Hall, Inc., 1949).

*Theory of statistics:* (1) Wilfred J. Dixon and Frank J. Massey, Jr., *Introduction to Statistical Analysis* (New York: McGraw-Hill Book Company, Inc., 1951). (2) Paul G. Hoel, *Introduction to Mathematical Statistics* (New York: John Wiley & Sons, Inc., 1954). (3) Helen M. Walker and Joseph Lev, *Statistical Inference* (New York: Henry Holt and Company, Inc., 1953).

*Experimental design:* Allen L. Edwards, *Experimental Design in Psychological Research* (New York: Rinehart & Company, Inc., 1950).

*Qualitative statistics:* G. Udny Yule and M. G. Kendall, *An Introduction to the Theory of Statistics* (New York: Hafner Publishing Company, 1950), pp. 1–68.

*Graphic techniques:* Calvin F. Schmid, *Handbook of Graphic Presentation* (New York: The Ronald Press Company, 1954).

*Problems of application to particular fields:* (1) Margaret Jarman Hagood and Daniel O. Price, *Statistics for Sociologists* (New York: Henry Holt and Company, Inc., 1952). (2) Quinn McNemar, *Psychological Statistics* (New York: John Wiley & Sons, Inc., 1955). (3) (For economics) Frederick E. Croxton and Dudley J. Cowden, *Applied General Statistics* (New York: Prentice Hall, Inc., 1945).

# 20 And a Few Grains of Salt

A major theme of this text is that statistics is not a substitute for thinking. No amount of statistical sophistication can overcome the deficiencies of a poorly designed experiment or support an illogical conclusion. This chapter consists of examples of misuses of statistics—errors of logic rather than errors of technique.

1. Since $4.20 per $100 of national income was spent on alcoholic beverages in France, compared with $.40 per $100 in the United States, a research agency concluded that Frenchmen drink ten times as much as Americans. Although the figures do indicate that Frenchmen spend a higher proportion of their income on alcohol, it does not show that they drink ten times as much. The national income of the United States is several times higher than in France, so that an equivalent amount of money is a smaller proportion of the national income in the United States. Other uncontrolled factors are the price of alcoholic beverages and the per cent of alcohol in the beverage. The evidence presented is not adequate to draw any conclusions concerning the extent to which Frenchmen drink more than Americans.

2. An American naturalist stated that a 17-year locust is a prophet of war. In 11 out of its last 12 appearances, from 1752 to 1939, a war began in the same year, said the report. Examination of the data shows a very loose definition of success and failure. For example, the Cherokee and Creek Indian Wars of 1837 are cited as instances of a successful prophecy. If such small and unimportant wars are included, the researcher would have little difficulty in finding the outbreak of a war in almost any year. Furthermore, the naturalist defined as a success the appearance

of the locust in 1752 which was followed by the outbreak of the
Seven Years' War in 1756. He failed to be consistent in his cri-
teria of success and failure.

3. A newspaper was enthusiastic about the decline of delinquency in Spokane County, Washington. In 1952 there were
1,072 cases, while there were only 1,000 cases in 1953. In the
same news story it was estimated that 2.3 per cent of the juvenile population was called to the attention of the court, with an
additional estimated 8.0 per cent of the juvenile population delinquent, but unreported. With so large a margin of error, the
slight decline in cases reported is obviously not worthy of
notice.

4. An advertisement boasted that each year found an increase
in the proportion of college graduates registering at a certain
New York hotel. Since there has been a steady increase in college
enrollment, this statistical analysis only proves that college
graduates may sleep in hotels.

5. A newspaper columnist noted that Presidents of the United
States lived to an average of 68 years, while sports figures died
at an average age of 61. In addition to the problem of compara-
bility of death rates from the eighteenth century to the present,
the columnist ignores the long apprenticeship necessary before
being elected President. A sports figure can become famous at
20, but Presidents must have lived to at least 40 before nomina-
tion to the Presidency.

6. A "psychologist" believed that sleeping from north to
south in bed was preferable because the person was in harmony
with magnetic currents. He wrote a letter to a newspaper asking
if others had had the same experience. He considered the 200
affirmative replies he received as confirming his hypothesis, ignoring the biased nature of this sample. By asking for only sub-
stantiating evidence, it would be difficult indeed to have dis-
proved his hypothesis.

7. A study of males with peptic ulcers showed that all of them
needed love and affection. No attempt was made to find how
many males without peptic ulcers needed love and affection.

8. A newspaper editorial said that senators and representa-
tives from the state of Washington are much more diligent and

faithful than members of the Congress from Eastern states, since they have an almost perfect attendance record in voting on crucial issues. Eastern congressmen are alleged to be more care-less in performing their duties since they miss many roll calls. An alternative explanation might be that congressmen from the Far West are unable to make frequent trips to their home dis-tricts by virtue of the time and distance involved.

9. Often newspapers attempt to rate the performance of Con-gress in terms of the proportion of bills requested by the admin-istration which were passed. Unfortunately for statistical analy-sis, administrations differ in the proportion of bills which are seriously proposed for passage. Some Presidents, faced with a hostile Congress, may prepare for an attack on the legislative branch by recommending a host of bills which have no chance of success.

10. A study found that 77.2 per cent of student husbands regularly help with the housework, while only 26.2 per cent of older married men help at home. The conclusion was made that certain values pertaining to the American family have changed in the last generation. The data reported actually do not bear out this contention. It is possible that the older men helped at home when they were young. It is also conceivable that the young students have working wives who need more help at home.

11. A large-scale study estimated that 10 per cent of the nation's public-school pupils are emotionally disturbed and need mental guidance. The survey was based on 2,540,888 cases in 350 school systems in 48 states. These data represent mere opinions taken from questionnaires submitted to a large sam-ple of school officials. The 10 per cent is therefore the mean of the collective judgments of these officials. Although ex-pressed statistically, this information is basically superficial and impressionistic.

12. An official of the Community Chest was defending the failure of social agencies to locate facilities in areas of low socio-economic status. He pointed out that the agencies needed volun-teer leaders and that the persons in these areas were not volun-teering in sufficient numbers. "In fact, we did a statistical study of this problem. There is a very high positive correlation between

median education and the proportion of volunteer leaders in an area. We'll just have to wait for these people to get educated."

The Community Chest official was confusing correlation and causation. Since education is an index of socioeconomic status, he was measuring the already known fact that the poorer areas were not providing a sufficient number of volunteer leaders. The correlation also would have been high between median rent and proportion of volunteers, but the official would not have thought of saying, "Let's wait until these people pay higher rents."

13. Statistical charts are often drawn in such a manner as to create distortion and misinterpretation. One common error is omission of the zero base line in a simple column chart. The results of such a practice are illustrated in Fig. 20-1. The graph in the upper panel (A) is drawn correctly with a scale running from 0 to 7,000. The graph in the bottom panel (B) has a vertical scale ranging from 5,000 to 7,000. The differences in the relative size of the columns give a grossly exaggerated and distorted impression. In (B), the height of the largest column is approximately eleven times that of the smallest. Actually, the difference between the minimum and maximum values is a little less than one and one-half scale intervals on the correctly drawn chart.

Sometimes it may be proper to omit the zero base line in a simple rectangular coordinate line chart. If this is done, however, there should be a clear indication of a break in the vertical scale.

Radically different impressions may be conveyed as the result of variations in the proportions of a rectangular coordinate chart. Figure 20-2 depicts a series of data presented on three grids of different proportions. The scales in grid A show good balance, whereas in both grids B and C one scale has been compressed and the other extended. As a consequence, the form of the curve in each instance has been greatly altered. Although it is extremely difficult, if not impossible, to specify hard-and-fast rules for determining the most appropriate proportion for any given chart, it is obvious that extremely contracted or extended scales should be avoided.

## OMISSION OF ZERO BASE LINE
## IN A COLUMN CHART

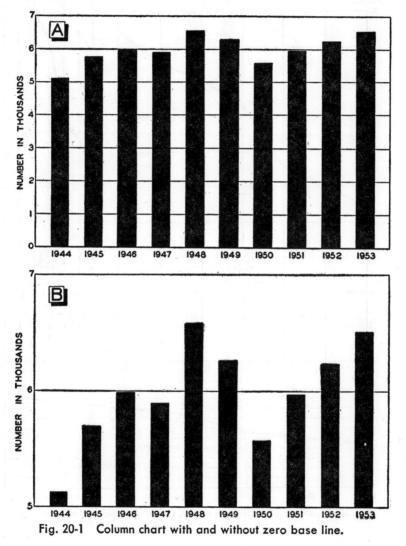

Fig. 20-1   Column chart with and without zero base line.

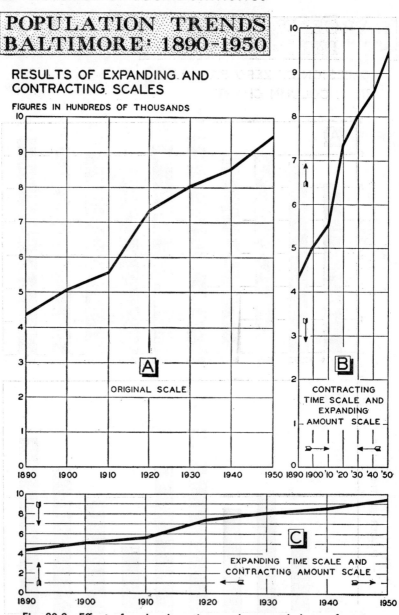

Fig. 20-2   Effect of scale alteration on shape and slope of curve.

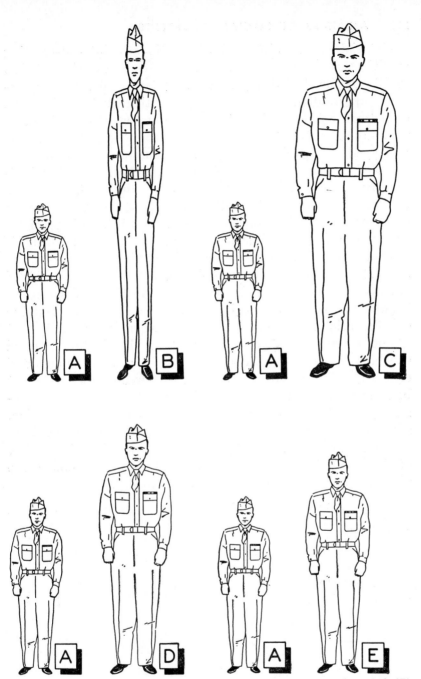

**Fig. 20-3   Comparison of sizes of pictorial symbols.** (*A*) original symbol; (*B*) original symbol doubled in height only; (*C*) original symbol doubled in both height and width; (*D*) symbol made twice the area of original symbol; (*E*) symbol made twice the volume of original symbol.

Pictorial charts with their dramatic and interest-creating qualities possess wide popular appeal. Since they are accepted so readily by the statistically unsophisticated public, it is particularly important that they be designed with great care. One type of pictorial chart that is widely used, but basically illogical and misleading, has the size of the symbols proportional to the values portrayed. For example, Figure 20-3 illustrates the difficulties involved in comparing pictorial symbols. If one wishes to indicate by means of pictorial symbols of this kind that the magnitude of one category is twice that of another, one symbol can be made twice the height or length of the other. This is precisely what would be done if a bar or column chart were used. The basis of comparison in this instance would be linear or one-dimensional. Suppose this logic were followed in comparing the size of two symbols representing human beings. In Fig. 20-3 symbol $B$ is twice the height and twice the area of symbol $A$, and the one-dimensional basis of comparison is logically acceptable. But symbol $B$ represents a grotesquely elongated caricature. Symbol $C$, which is artistically proportioned, is four times the area and eight times the volume of $A$. Under the circumstances it might be assumed that the proper solution is to reconstruct the symbols on the basis of a more consistent and comparable criterion, such as area or volume. Symbol $D$ represents twice the area of $A$, and symbol $E$ twice the volume. However, from the point of view of visual impression, comparisons of $A$ and $D$ and $A$ and $E$ are virtually incomprehensible with respect to size.

# Appendix

## Explanation of Table A

For example, to square a three-digit number, find the first two digits in the extreme left-hand column labeled $N$. The third digit of the number you wish to square is represented by the caption of a column. Move down that column until the row in column $N$ is reached which contains the first two digits. At that point you have the square of your number.

When there are decimals, certain adjustments have to be made. For example, the square of .384, 3.84, 38.4, 384, and 3,840, respectively, are given as the same number in the table—147,456. To determine the position of the decimal, round your original number and note the approximate square. For example, 38.4 is close to 40. The square of 40 is 1,600, and the square of 38.4 must be of the same order of magnitude. Accordingly, the square of 38.4 must be 1,474.56.

To find the square root of a number, merely reverse the process. Start by finding your number in the body of the table, and then read off the square root from the heading of that row and column. There is one major caution to be observed. The square root of 9 is not the same as the square root of 90. Each number can be found in two places in the body of the table. All that is needed is an approximate idea of the correct answer. The square root of 9.17 is about 3.03, and the square root of 91.7 is approximately 9.58.

## Table A   Squares and Square Roots of Numbers*

| N | 0 | 1 | 2 | 3 | 4 | 5 | 6 | 7 | 8 | 9 |
|---|---|---|---|---|---|---|---|---|---|---|
| 100 | 10000 | 10201 | 10404 | 10609 | 10816 | 11025 | 11236 | 11449 | 11664 | 11881 |
| 110 | 12100 | 12321 | 12544 | 12769 | 12996 | 13225 | 13456 | 13689 | 13924 | 14161 |
| 120 | 14400 | 14641 | 14884 | 15129 | 15376 | 15625 | 15876 | 16129 | 16384 | 16641 |
| 130 | 16900 | 17161 | 17424 | 17689 | 17956 | 18225 | 18496 | 18769 | 19044 | 19321 |
| 140 | 19600 | 19881 | 20164 | 20449 | 20736 | 21025 | 21316 | 21609 | 21904 | 22201 |
| 150 | 22500 | 22801 | 23104 | 23409 | 23716 | 24025 | 24336 | 24649 | 24964 | 25281 |
| 160 | 25600 | 25921 | 26244 | 26569 | 26896 | 27225 | 27556 | 27889 | 28224 | 28561 |
| 170 | 28900 | 29241 | 29584 | 29929 | 30276 | 30625 | 30976 | 31329 | 31684 | 32041 |
| 180 | 32400 | 32761 | 33124 | 33489 | 33856 | 34225 | 34596 | 34969 | 35344 | 35721 |
| 190 | 36100 | 36481 | 36864 | 37249 | 37636 | 38025 | 38416 | 38809 | 39204 | 39601 |
| 200 | 40000 | 40401 | 40804 | 41209 | 41616 | 42025 | 42436 | 42849 | 43264 | 43681 |
| 210 | 44100 | 44521 | 44944 | 45369 | 45796 | 46225 | 46656 | 47089 | 47524 | 47961 |
| 220 | 48400 | 48841 | 49284 | 49729 | 50176 | 50625 | 51076 | 51529 | 51984 | 52441 |
| 230 | 52900 | 53361 | 53824 | 54289 | 54756 | 55225 | 55696 | 56169 | 56644 | 57121 |
| 240 | 57600 | 58081 | 58564 | 59049 | 59536 | 60025 | 60516 | 61009 | 61504 | 62001 |
| 250 | 62500 | 63001 | 63504 | 64009 | 64516 | 65025 | 65536 | 66049 | 66564 | 67081 |
| 260 | 67600 | 68121 | 68644 | 69169 | 69696 | 70225 | 70756 | 71289 | 71824 | 72361 |
| 270 | 72900 | 73441 | 73984 | 74529 | 75076 | 75625 | 76176 | 76729 | 77284 | 77841 |
| 280 | 78400 | 78961 | 79524 | 80089 | 80656 | 81225 | 81796 | 82369 | 82944 | 83521 |
| 290 | 84100 | 84681 | 85264 | 85849 | 86436 | 87025 | 87616 | 88209 | 88804 | 89401 |
| 300 | 90000 | 90601 | 91204 | 91809 | 92416 | 93025 | 93636 | 94249 | 94864 | 95481 |
| 310 | 96100 | 96721 | 97344 | 97969 | 98596 | 99225 | 99856 | 100489 | 101124 | 101761 |
| 320 | 102400 | 103041 | 103684 | 104329 | 104976 | 105625 | 106276 | 106929 | 107584 | 108241 |
| 330 | 108900 | 109561 | 110224 | 110889 | 111556 | 112225 | 112896 | 113569 | 114244 | 114921 |
| 340 | 115600 | 116281 | 116964 | 117649 | 118336 | 119025 | 119716 | 120409 | 121104 | 121801 |
| 350 | 122500 | 123201 | 123904 | 124609 | 125316 | 126025 | 126736 | 127449 | 128164 | 128881 |
| 360 | 129600 | 130321 | 131044 | 131769 | 132496 | 133225 | 133956 | 134689 | 135424 | 136161 |
| 370 | 136900 | 137641 | 138384 | 139129 | 139876 | 140625 | 141376 | 142129 | 142884 | 143641 |
| 380 | 144400 | 145161 | 145924 | 146689 | 147456 | 148225 | 148996 | 149769 | 150544 | 151321 |
| 390 | 152100 | 152881 | 153664 | 154449 | 155236 | 156025 | 156816 | 157609 | 158404 | 159201 |
| 400 | 160000 | 160801 | 161604 | 162409 | 163216 | 164025 | 164836 | 165649 | 166464 | 167281 |
| 410 | 168100 | 168921 | 169744 | 170569 | 171396 | 172225 | 173056 | 173889 | 174724 | 175561 |
| 420 | 176400 | 177241 | 178084 | 178929 | 179776 | 180625 | 181476 | 182329 | 183184 | 184041 |
| 430 | 184900 | 185761 | 186624 | 187489 | 188356 | 189225 | 190096 | 190969 | 191844 | 192721 |
| 440 | 193600 | 194481 | 195364 | 196249 | 197136 | 198025 | 198916 | 199809 | 200704 | 201601 |
| 450 | 202500 | 203401 | 204304 | 205209 | 206116 | 207025 | 207936 | 208849 | 209764 | 210681 |
| 460 | 211600 | 212521 | 213444 | 214369 | 215296 | 216225 | 217156 | 218089 | 219024 | 219961 |
| 470 | 220900 | 221841 | 222784 | 223729 | 224676 | 225625 | 226576 | 227529 | 228484 | 229441 |
| 480 | 230400 | 231361 | 232324 | 233289 | 234256 | 235225 | 236196 | 237169 | 238144 | 239121 |
| 490 | 240100 | 241081 | 242064 | 243049 | 244036 | 245025 | 246016 | 247009 | 248004 | 249001 |
| 500 | 250000 | 251001 | 252004 | 253009 | 254016 | 255025 | 256036 | 257049 | 258064 | 259081 |
| 510 | 260100 | 261121 | 262144 | 263169 | 264196 | 265225 | 266256 | 267289 | 268324 | 269361 |
| 520 | 270400 | 271441 | 272484 | 273529 | 274576 | 275625 | 276676 | 277729 | 278784 | 279841 |
| 530 | 280900 | 281961 | 283024 | 284089 | 285156 | 286225 | 287296 | 288369 | 289444 | 290521 |
| 540 | 291600 | 292681 | 293764 | 294849 | 295936 | 297025 | 298116 | 299209 | 300304 | 301401 |

* By permission from *Statistical Tables and Problems*, by A. E. Waugh. Copyright, 1952, McGraw-Hill Book Company, Inc.

Table A Squares and Square Roots of Numbers (*Continued*)

| N | 0 | 1 | 2 | 3 | 4 | 5 | 6 | 7 | 8 | 9 |
|---|---|---|---|---|---|---|---|---|---|---|
| 550 | 302500 | 303601 | 304704 | 305809 | 306916 | 308025 | 309136 | 310249 | 311364 | 312481 |
| 560 | 313600 | 314721 | 315844 | 316969 | 318096 | 319225 | 320356 | 321489 | 322624 | 323761 |
| 570 | 324900 | 326041 | 327184 | 328329 | 329476 | 330625 | 331776 | 332929 | 334084 | 335241 |
| 580 | 336400 | 337561 | 338724 | 339889 | 341056 | 342225 | 343396 | 344569 | 345744 | 346921 |
| 590 | 348100 | 349281 | 350464 | 351649 | 352836 | 354025 | 355216 | 356409 | 357604 | 358801 |
| 600 | 360000 | 361201 | 362404 | 363609 | 364816 | 366025 | 367236 | 368449 | 369664 | 370881 |
| 610 | 372100 | 373321 | 374544 | 375769 | 376996 | 378225 | 379456 | 380689 | 381924 | 383161 |
| 620 | 384400 | 385641 | 386884 | 388129 | 389376 | 390625 | 391876 | 393129 | 394384 | 395641 |
| 630 | 396900 | 398161 | 399424 | 400689 | 401956 | 403225 | 404496 | 405769 | 407044 | 408321 |
| 640 | 409600 | 410881 | 412164 | 413449 | 414736 | 416025 | 417316 | 418609 | 419904 | 421201 |
| 650 | 422500 | 423801 | 425104 | 426409 | 427716 | 429025 | 430336 | 431649 | 432964 | 434281 |
| 660 | 435600 | 436921 | 438244 | 439569 | 440896 | 442225 | 443556 | 444889 | 446224 | 447561 |
| 670 | 448900 | 450241 | 451584 | 452929 | 454276 | 455625 | 456976 | 458329 | 459684 | 461041 |
| 680 | 462400 | 463761 | 465124 | 466489 | 467856 | 469225 | 470596 | 471969 | 473344 | 474721 |
| 690 | 476100 | 477481 | 478864 | 480249 | 481636 | 483025 | 484416 | 485809 | 487204 | 488601 |
| 700 | 490000 | 491401 | 492804 | 494209 | 495616 | 497025 | 498436 | 498849 | 501264 | 502681 |
| 710 | 504100 | 505521 | 506944 | 508369 | 509796 | 511225 | 512656 | 514089 | 515524 | 516961 |
| 720 | 518400 | 519841 | 521284 | 522729 | 524176 | 525625 | 527076 | 528529 | 529984 | 531441 |
| 730 | 532900 | 534361 | 535824 | 537289 | 538756 | 540225 | 541696 | 543169 | 544644 | 546121 |
| 740 | 547600 | 549081 | 550564 | 552049 | 553536 | 555025 | 556516 | 558009 | 559504 | 561001 |
| 750 | 562500 | 564001 | 565504 | 567009 | 568516 | 570025 | 571536 | 573049 | 574564 | 576081 |
| 760 | 577600 | 579121 | 580644 | 582169 | 583696 | 585225 | 586756 | 588289 | 589824 | 591361 |
| 770 | 592900 | 594441 | 595984 | 597529 | 599076 | 600625 | 602176 | 603729 | 605284 | 606841 |
| 780 | 608400 | 609961 | 611524 | 613089 | 614656 | 616225 | 617796 | 619369 | 620944 | 622521 |
| 790 | 624100 | 625681 | 627264 | 628849 | 630436 | 632025 | 633616 | 635209 | 636804 | 638401 |
| 800 | 640000 | 641601 | 643204 | 644809 | 646416 | 648025 | 649636 | 651249 | 652864 | 654481 |
| 810 | 656100 | 657721 | 659344 | 660969 | 662596 | 664225 | 665856 | 667489 | 669124 | 670761 |
| 820 | 672400 | 674041 | 675684 | 677329 | 678976 | 680625 | 682276 | 683929 | 685584 | 687241 |
| 830 | 688900 | 690561 | 692224 | 693889 | 695556 | 697225 | 698896 | 700569 | 702244 | 703921 |
| 840 | 705600 | 707281 | 708964 | 710649 | 712336 | 714025 | 715716 | 717409 | 719104 | 720801 |
| 850 | 722500 | 724201 | 725904 | 727609 | 729316 | 731025 | 732736 | 734449 | 736164 | 737881 |
| 860 | 739600 | 741321 | 743044 | 744769 | 746496 | 748225 | 749956 | 751689 | 753424 | 755161 |
| 870 | 756900 | 758641 | 760384 | 762129 | 763876 | 765625 | 767376 | 769129 | 770884 | 772641 |
| 880 | 774400 | 776161 | 777924 | 779689 | 781456 | 783225 | 784996 | 786769 | 788544 | 790321 |
| 890 | 792100 | 793881 | 795664 | 797449 | 799236 | 801025 | 802816 | 804609 | 806404 | 808201 |
| 900 | 810000 | 811801 | 813604 | 815409 | 817216 | 819025 | 820836 | 822649 | 824464 | 826281 |
| 910 | 828100 | 829921 | 831744 | 833569 | 835396 | 837225 | 839056 | 840889 | 842724 | 844561 |
| 920 | 846400 | 848241 | 850084 | 851929 | 853776 | 855625 | 857476 | 859329 | 861184 | 863041 |
| 930 | 864900 | 866761 | 868624 | 870489 | 872356 | 874225 | 876096 | 877969 | 879844 | 881721 |
| 940 | 883600 | 885481 | 887364 | 889249 | 891136 | 893025 | 894916 | 896809 | 898704 | 900601 |
| 950 | 902500 | 904401 | 906304 | 908209 | 910116 | 912025 | 913936 | 915849 | 917764 | 919681 |
| 960 | 921600 | 923521 | 925444 | 927369 | 929296 | 931225 | 933156 | 935089 | 937024 | 938961 |
| 970 | 940900 | 942841 | 944784 | 946729 | 948676 | 950625 | 952576 | 954529 | 956484 | 958441 |
| 980 | 960400 | 962361 | 964324 | 966289 | 968256 | 970225 | 972196 | 974169 | 976144 | 978121 |
| 990 | 980100 | 982081 | 984064 | 986049 | 988036 | 990025 | 992016 | 994009 | 996004 | 998001 |

## Table B    Areas under the Normal Curve*

Fractional parts of the total area (1.000) under the normal curve between the mean and a perpendicular erected at various numbers of standard deviations ($x/\sigma$) from the mean. To illustrate the use of the table, 39.065 per cent of the total area under the curve will lie between the mean and a perpendicular erected at a distance of $1.23\sigma$ from the mean.

Each figure in the body of the table is preceded by a decimal point.

| $x/\sigma$ | .00 | .01 | .02 | .03 | .04 | .05 | .06 | .07 | .08 | .09 |
|---|---|---|---|---|---|---|---|---|---|---|
| 0.0 | 00000 | 00399 | 00798 | 01197 | 01595 | 01994 | 02392 | 02790 | 03188 | 03586 |
| 0.1 | 03983 | 04380 | 04776 | 05172 | 05567 | 05962 | 06356 | 06749 | 07142 | 07535 |
| 0.2 | 07926 | 08317 | 08706 | 09095 | 09483 | 09871 | 10257 | 10642 | 11026 | 11409 |
| 0.3 | 11791 | 12172 | 12552 | 12930 | 13307 | 13683 | 14058 | 14431 | 14803 | 15173 |
| 0.4 | 15554 | 15910 | 16276 | 16640 | 17003 | 17364 | 17724 | 18082 | 18439 | 18793 |
| 0.5 | 19146 | 19497 | 19847 | 20194 | 20450 | 20884 | 21226 | 21566 | 21904 | 22240 |
| 0.6 | 22575 | 22907 | 23237 | 23565 | 23891 | 24215 | 24537 | 24857 | 25175 | 25490 |
| 0.7 | 25804 | 26115 | 26424 | 26730 | 27035 | 27337 | 27637 | 27935 | 28230 | 28524 |
| 0.8 | 28814 | 29103 | 29389 | 29673 | 29955 | 30234 | 30511 | 30785 | 31057 | 31327 |
| 0.9 | 31594 | 31859 | 32121 | 32381 | 32639 | 32894 | 33147 | 33398 | 33646 | 33891 |
| 1.0 | 34134 | 34375 | 34614 | 34850 | 35083 | 35313 | 35543 | 35769 | 35993 | 36214 |
| 1.1 | 36433 | 36650 | 36864 | 37076 | 37286 | 37493 | 37698 | 37900 | 38100 | 38298 |
| 1.2 | 38493 | 38686 | 38877 | 39065 | 39251 | 39435 | 39617 | 39796 | 39973 | 40147 |
| 1.3 | 40320 | 40490 | 40658 | 40824 | 40988 | 41149 | 41308 | 41466 | 41621 | 41774 |
| 1.4 | 41924 | 42073 | 42220 | 42364 | 42507 | 42647 | 42786 | 42922 | 43056 | 43189 |
| 1.5 | 43319 | 43448 | 43574 | 43699 | 43822 | 43943 | 44062 | 44179 | 44295 | 44408 |
| 1.6 | 44520 | 44630 | 44738 | 44845 | 44950 | 45053 | 45154 | 45254 | 45352 | 45449 |
| 1.7 | 45543 | 45637 | 45728 | 45818 | 45907 | 45994 | 46080 | 46164 | 46246 | 46327 |
| 1.8 | 46407 | 46485 | 46562 | 46638 | 46712 | 46784 | 46856 | 46926 | 46995 | 47062 |
| 1.9 | 47128 | 47193 | 47257 | 47320 | 47381 | 47441 | 47500 | 47558 | 47615 | 47670 |
| 2.0 | 47725 | 47778 | 47831 | 47882 | 47932 | 47982 | 48030 | 48077 | 48124 | 48169 |
| 2.1 | 48214 | 48257 | 48300 | 48341 | 48382 | 48422 | 48461 | 48500 | 48537 | 48574 |
| 2.2 | 48610 | 48645 | 48679 | 48713 | 48745 | 48778 | 48809 | 48840 | 48870 | 48899 |
| 2.3 | 48928 | 48956 | 48983 | 49010 | 49036 | 49061 | 49086 | 49111 | 49134 | 49158 |
| 2.4 | 49180 | 49202 | 49224 | 49245 | 49266 | 49286 | 49305 | 49324 | 49343 | 49361 |
| 2.5 | 49379 | 49396 | 49413 | 49430 | 49446 | 49461 | 49477 | 49492 | 49506 | 49520 |
| 2.6 | 49534 | 49547 | 49560 | 49573 | 49585 | 49598 | 49609 | 49621 | 49632 | 49643 |
| 2.7 | 49653 | 49664 | 49674 | 49683 | 49693 | 49702 | 49711 | 49720 | 49728 | 49736 |
| 2.8 | 49744 | 49752 | 49760 | 49767 | 49774 | 49781 | 49788 | 49795 | 49801 | 49807 |
| 2.9 | 49813 | 49819 | 49825 | 49831 | 49836 | 49841 | 49846 | 49851 | 49856 | 49861 |
| 3.0 | 49865 | | | | | | | | | |
| 3.5 | 4997674 | | | | | | | | | |
| 4.0 | 4999683 | | | | | | | | | |
| 4.5 | 4999966 | | | | | | | | | |
| 5.0 | 4999997133 | | | | | | | | | |

* Reprinted from *Statistical Tables and Problems*, by A. E. Waugh, copyright, 1952; adapted from *Elements of Statistics*, by F. C. Kent, copyright, 1924, McGraw-Hill Book Company, Inc.

## Table C  Values of $t^*$

| d.f. | $t_{.95}$ | $t_{.975}$ | $t_{.9875}$ | $t_{.995}$ | $t_{.9975}$ |
|------|------|------|------|------|------|
| 1 | 6.31 | 12.7 | 25.5 | 63.7 | 127 |
| 2 | 2.92 | 4.30 | 6.21 | 9.92 | 14.1 |
| 3 | 2.35 | 3.18 | 4.18 | 5.84 | 7.45 |
| 4 | 2.13 | 2.78 | 3.50 | 4.60 | 5.60 |
| 5 | 2.01 | 2.57 | 3.16 | 4.03 | 4.77 |
| 6 | 1.94 | 2.45 | 2.97 | 3.71 | 4.32 |
| 7 | 1.89 | 2.36 | 2.84 | 3.50 | 4.03 |
| 8 | 1.86 | 2.31 | 2.75 | 3.36 | 3.83 |
| 9 | 1.83 | 2.26 | 2.69 | 3.25 | 3.69 |
| 10 | 1.81 | 2.23 | 2.63 | 3.17 | 3.58 |
| 11 | 1.80 | 2.20 | 2.59 | 3.11 | 3.50 |
| 12 | 1.78 | 2.18 | 2.56 | 3.05 | 3.43 |
| 13 | 1.77 | 2.16 | 2.53 | 3.01 | 3.37 |
| 14 | 1.76 | 2.14 | 2.51 | 2.98 | 3.33 |
| 15 | 1.75 | 2.13 | 2.49 | 2.95 | 3.29 |
| 16 | 1.75 | 2.12 | 2.47 | 2.92 | 3.25 |
| 17 | 1.74 | 2.11 | 2.46 | 2.90 | 3.22 |
| 18 | 1.73 | 2.10 | 2.45 | 2.88 | 3.20 |
| 19 | 1.73 | 2.09 | 2.43 | 2.86 | 3.17 |
| 20 | 1.72 | 2.09 | 2.42 | 2.85 | 3.15 |
| 21 | 1.72 | 2.08 | 2.41 | 2.83 | 3.14 |
| 22 | 1.72 | 2.07 | 2.41 | 2.82 | 3.12 |
| 23 | 1.71 | 2.07 | 2.40 | 2.81 | 3.10 |
| 24 | 1.71 | 2.06 | 2.39 | 2.80 | 3.09 |
| 25 | 1.71 | 2.06 | 2.38 | 2.79 | 3.08 |
| 26 | 1.71 | 2.06 | 2.38 | 2.78 | 3.07 |
| 27 | 1.70 | 2.05 | 2.37 | 2.77 | 3.06 |
| 28 | 1.70 | 2.05 | 2.37 | 2.76 | 3.05 |
| 29 | 1.70 | 2.05 | 2.36 | 2.76 | 3.04 |
| 30 | 1.70 | 2.04 | 2.36 | 2.75 | 3.03 |
| 40 | **1.68** | 2.02 | 2.33 | 2.70 | 2.97 |
| 60 | 1.67 | 2.00 | 2.30 | 2.66 | 2.91 |
| 120 | 1.66 | 1.98 | 2.27 | 2.62 | 2.86 |
| ∞ | 1.64 | 1.96 | 2.24 | 2.58 | 2.81 |
| d.f. | $t_{.05}$ | $t_{.025}$ | $t_{.0125}$ | $t_{.005}$ | $t_{.0025}$ |

When the table is read from the foot, the tabled values are to be prefixed with a negative sign. Interpolation should be performed using the reciprocals of the degrees of freedom.

The values in the above table were computed from percentiles of the $F$ distribution.

* By permission from *Introduction to Statistical Analysis*, by W. J. Dixon and F. J. Massey, Jr. Copyright, 1951, McGraw-Hill Book Company, Inc.

### Table D    Values of $r$ at the 5 and 1 per cent Levels of Significance

| Degrees of freedom | 5 per cent | 1 per cent | Degrees of freedom | 5 per cent | 1 per cent |
|:---:|:---:|:---:|:---:|:---:|:---:|
| 1 | .997 | 1.000 | 24 | .388 | .496 |
| 2 | .950 | .990 | 25 | .381 | .487 |
| 3 | .878 | .959 | 26 | .374 | .478 |
| 4 | .811 | .917 | 27 | .367 | .470 |
| 5 | .754 | .874 | 28 | .361 | .463 |
| 6 | .707 | .834 | 29 | .355 | .456 |
| 7 | .666 | .798 | 30 | .349 | .449 |
| 8 | .632 | .765 | 35 | .325 | .418 |
| 9 | .602 | .735 | 40 | .304 | .393 |
| 10 | .576 | .708 | 45 | .288 | .372 |
| 11 | .553 | .684 | 50 | .273 | .354 |
| 12 | .532 | .661 | 60 | .250 | .325 |
| 13 | .514 | .641 | 70 | .232 | .302 |
| 14 | .497 | .623 | 80 | .217 | .283 |
| 15 | .482 | .606 | 90 | .205 | .267 |
| 16 | .468 | .590 | 100 | .195 | .254 |
| 17 | .456 | .575 | 125 | .174 | .228 |
| 18 | .444 | .561 | 150 | .159 | .208 |
| 19 | .433 | .549 | 200 | .138 | .181 |
| 20 | .423 | .537 | 300 | .113 | .148 |
| 21 | .413 | .526 | 400 | .098 | .128 |
| 22 | .404 | .515 | 500 | .088 | .115 |
| 23 | .396 | .505 | 1000 | .062 | .081 |

This table taken from George W. Snedecor, *Statistical Methods*, Ames, Iowa: The Iowa State College Press, 1946, p. 149. Portions of this table are reprinted from Table V.A of R. A. Fisher: *Statistical Methods for Research Workers*, published by Oliver and Boyd Ltd., Edinburgh, by permission of the author and publishers.

## Table E   Values of $Z$ for Given Values of $r$*

| r | .000 | .001 | .002 | .003 | .004 | .005 | .006 | .007 | .008 | .009 |
|---|------|------|------|------|------|------|------|------|------|------|
| .000 | .0000 | .0010 | .0020 | .0030 | .0040 | .0050 | .0060 | .0070 | .0080 | .0090 |
| .010 | .0100 | .0110 | .0120 | .0130 | .0140 | .0150 | .0160 | .0170 | .0180 | .0190 |
| .020 | .0200 | .0210 | .0220 | .0230 | .0240 | .0250 | .0260 | .0270 | .0280 | .0290 |
| .030 | .0300 | .0310 | .0320 | .0330 | .0340 | .0350 | .0360 | .0370 | .0380 | .0390 |
| .040 | .0400 | .0410 | .0420 | .0430 | .0440 | .0450 | .0460 | .0470 | .0480 | .0490 |
| .050 | .0501 | .0511 | .0521 | .0531 | .0541 | .0551 | .0561 | .0571 | .0581 | .0591 |
| .060 | .0601 | .0611 | .0621 | .0631 | .0641 | .0651 | .0661 | .0671 | .0681 | .0691 |
| .070 | .0701 | .0711 | .0721 | .0731 | .0741 | .0751 | .0761 | .0771 | .0782 | .0792 |
| .080 | .0802 | .0812 | .0822 | .0832 | .0842 | .0852 | .0862 | .0872 | .0882 | .0892 |
| .090 | .0902 | .0912 | .0922 | .0933 | .0943 | .0953 | .0963 | .0973 | .0983 | .0993 |
| .100 | .1003 | .1013 | .1024 | .1034 | .1044 | .1054 | .1064 | .1074 | .1084 | .1094 |
| .110 | .1105 | .1115 | .1125 | .1135 | .1145 | .1155 | .1165 | .1175 | .1185 | .1195 |
| .120 | .1206 | .1216 | .1226 | .1236 | .1246 | .1257 | .1267 | .1277 | .1287 | .1297 |
| .130 | .1308 | .1318 | .1328 | .1338 | .1348 | .1358 | .1368 | .1379 | .1389 | .1399 |
| .140 | .1409 | .1419 | .1430 | .1440 | .1450 | .1460 | .1470 | .1481 | .1491 | .1501 |
| .150 | .1511 | .1522 | .1532 | .1542 | .1552 | .1563 | .1573 | .1583 | .1593 | .1604 |
| .160 | .1614 | .1624 | .1634 | .1654 | .1655 | .1665 | .1676 | .1686 | .1696 | .1706 |
| .170 | .1717 | .1727 | .1737 | .1748 | .1758 | .1768 | .1779 | .1789 | .1799 | .1810 |
| .180 | .1820 | .1830 | .1841 | .1851 | .1861 | .1872 | .1882 | .1892 | .1903 | .1913 |
| .190 | .1923 | .1934 | .1944 | .1954 | .1965 | .1975 | .1986 | .1996 | .2007 | .2017 |
| .200 | .2027 | .2038 | .2048 | .2059 | .2069 | .2079 | .2090 | .2100 | .2111 | .2121 |
| .210 | .2132 | .2142 | .2153 | .2163 | .2174 | .2184 | .2194 | .2205 | .2215 | .2226 |
| .220 | .2237 | .2247 | .2258 | .2268 | .2279 | .2289 | .2300 | .2310 | .2321 | .2331 |
| .230 | .2342 | .2353 | .2363 | .2374 | .2384 | .2395 | .2405 | .2416 | .2427 | .2437 |
| .240 | .2448 | .2458 | .2469 | .2480 | .2490 | .2501 | .2511 | .2522 | .2533 | .2543 |
| .250 | .2554 | .2565 | .2575 | .2586 | .2597 | .2608 | .2618 | .2629 | .2640 | .2650 |
| .260 | .2661 | .2672 | .2682 | .2693 | .2704 | .2715 | .2726 | .2736 | .2747 | .2758 |
| .370 | .2769 | .2779 | .2790 | .2801 | .2812 | .2823 | .2833 | .2844 | .2855 | .2866 |
| .280 | .2877 | .2888 | .2898 | .2909 | .2920 | .2931 | .2942 | .2953 | .2964 | .2975 |
| .290 | .2986 | .2997 | .3008 | .3019 | .3029 | .3040 | .3051 | .3062 | .3073 | .3084 |
| .300 | .3095 | .3106 | .3117 | .3128 | .3139 | .3150 | .3161 | .3172 | .3183 | .3195 |
| .310 | .3206 | .3217 | .3228 | .3239 | .3250 | .3261 | .3272 | .3283 | .3294 | .3305 |
| .320 | .3317 | .3328 | .3339 | .3350 | .3361 | .3372 | .3384 | .3395 | .3406 | .3417 |
| .330 | .3428 | .3439 | .3451 | .3462 | .3473 | .3484 | .3496 | .3507 | .3518 | .3530 |
| .340 | .3541 | .3552 | .3564 | .3575 | .3586 | .3597 | .3609 | .3620 | .3632 | .3643 |
| .350 | .3654 | .3666 | .3677 | .3689 | .3700 | .3712 | .3723 | .3734 | .3746 | .3757 |
| .360 | .3769 | .3780 | .3792 | .3803 | .3815 | .3826 | .3838 | .3850 | .3861 | .3873 |
| .370 | .3884 | .3896 | .3907 | .3919 | .3931 | .3942 | .3954 | .3966 | .3977 | .3989 |
| .380 | .4001 | .4012 | .4024 | .4036 | .4047 | .4059 | .4071 | .4083 | .4094 | .4106 |
| .390 | .4118 | .4130 | .4142 | .4153 | .4165 | .4177 | .4189 | .4201 | .4213 | .4225 |
| .400 | .4236 | .4248 | .4260 | .4272 | .4284 | .4296 | .4308 | .4320 | .4332 | .4344 |
| .410 | .4356 | .4368 | .4380 | .4392 | .4404 | .4416 | .4429 | .4441 | .4453 | .4465 |
| .420 | .4477 | .4489 | .4501 | .4513 | .4526 | .4538 | .4550 | .4562 | .4574 | .4587 |
| .430 | .4599 | .4611 | .4623 | .4636 | .4648 | .4660 | .4673 | .4685 | .4697 | .4710 |
| .440 | .4722 | .4735 | .4747 | .4760 | .4772 | .4784 | .4797 | .4809 | .4822 | .4835 |
| .450 | .4847 | .4860 | .4872 | .4885 | .4897 | .4910 | .4923 | .4935 | .4948 | .4961 |
| .460 | .4973 | .4986 | .4999 | .5011 | .5024 | .5037 | .5049 | .5062 | .5075 | .5088 |
| .470 | .5101 | .5114 | .5126 | .5139 | .5152 | .5165 | .5178 | .5191 | .5204 | .5217 |
| .480 | .5230 | .5243 | .5256 | .5279 | .5282 | .5295 | .5308 | .5321 | .5334 | .5347 |
| .490 | .5361 | .5374 | .5387 | .5400 | .5413 | .5427 | .5440 | .5453 | .5466 | .5480 |

## Table E    Values of $Z$    (*Continued*)

| r | .000 | .001 | .002 | .003 | .004 | .005 | .006 | .007 | .008 | .009 |
|---|---|---|---|---|---|---|---|---|---|---|
| .500 | .5493 | .5506 | .5520 | .5533 | .5547 | .5560 | .5573 | .5587 | .5600 | .5614 |
| .510 | .5627 | .5641 | .5654 | .5668 | .5681 | .5695 | .5709 | .5722 | .5736 | .5750 |
| .520 | .5763 | .5777 | .5791 | .5805 | .5818 | .5832 | .5846 | .5860 | .5874 | .5888 |
| .530 | .5901 | .5915 | .5929 | .5943 | .5957 | .5971 | .5985 | .5999 | .6013 | .6027 |
| .540 | .6042 | .6056 | .6070 | .6084 | .6098 | .6112 | .6127 | .6141 | .6155 | .6170 |
| .550 | .6184 | .6198 | .6213 | .6227 | .6241 | .6256 | .6270 | .6285 | .6299 | .6314 |
| .560 | .6328 | .6343 | .6358 | .6372 | .6387 | .6401 | .6416 | .6431 | .6446 | .6460 |
| .570 | .6475 | .6490 | .6505 | .6520 | .6535 | .6550 | .6565 | .6579 | .6594 | .6610 |
| .580 | .6625 | .6640 | .6655 | .6670 | .6685 | .6700 | .6715 | .6731 | .6746 | .6761 |
| .590 | .6777 | .6792 | .6807 | .6823 | .6838 | .6854 | .6869 | .6885 | .6900 | .6916 |
| .600 | .6931 | .6947 | .6963 | .6978 | .6994 | .7010 | .7026 | .7042 | .7057 | .7073 |
| .610 | .7089 | .7105 | .7121 | .7137 | .7153 | .7169 | .7185 | .7201 | .7218 | .7234 |
| .620 | .7250 | .7266 | .7283 | .7299 | .7315 | .7332 | .7348 | .7364 | .7381 | .7398 |
| .630 | .7414 | .7431 | .7447 | .7464 | .7481 | .7497 | .7514 | .7531 | .7548 | .7565 |
| .640 | .7582 | .7599 | .7616 | .7633 | .7650 | .7667 | .7684 | .7701 | .7718 | .7736 |
| .650 | .7753 | .7770 | .7788 | .7805 | .7823 | .7840 | .7858 | .7875 | .7893 | .7910 |
| .660 | .7928 | .7946 | .7964 | .7981 | .7999 | .8017 | .8035 | .8053 | .8071 | .8089 |
| .670 | .8107 | .8126 | .8144 | .8162 | .8180 | .8199 | .8217 | .8236 | .8254 | .8273 |
| .680 | .8291 | .8310 | .8328 | .8347 | .8366 | .8385 | .8404 | .8423 | .8442 | .8461 |
| .690 | .8480 | .8499 | .8518 | .8537 | .8556 | .8576 | .8595 | .8614 | .8634 | .8653 |
| .700 | .8673 | .8693 | .8712 | .8732 | .8752 | .8772 | .8792 | .8812 | .8832 | .8852 |
| .710 | .8872 | .8892 | .8912 | .8933 | .8953 | .8973 | .8994 | .9014 | .9035 | .9056 |
| .720 | .9076 | .9097 | .9118 | .9139 | .9160 | .9181 | .9202 | .9223 | .9245 | .9266 |
| .730 | .9287 | .9309 | .9330 | .9352 | .9373 | .9395 | .9417 | .9439 | .9461 | .9483 |
| .740 | .9505 | .9527 | .9549 | .9571 | .9594 | .9616 | .9639 | .9661 | .9684 | .9707 |
| .750 | .9730 | .9752 | .9775 | .9799 | .9822 | .9845 | .9868 | .9892 | .9915 | .9939 |
| .760 | .9962 | .9986 | 1.0010 | 1.0034 | 1.0058 | 1.0082 | 1.0106 | 1.0130 | 1.0154 | 1.0179 |
| .770 | 1.0203 | 1.0228 | 1.0253 | 1.0277 | 1.0302 | 1.0327 | 1.0352 | 1.0378 | 1.0403 | 1.0428 |
| .780 | 1.0454 | 1.0479 | 1.0505 | 1.0531 | 1.0557 | 1.0583 | 1.0609 | 1.0635 | 1.0661 | 1.0688 |
| .790 | 1.0714 | 1.0741 | 1.0768 | 1.0795 | 1.0822 | 1.0849 | 1.0876 | 1.0903 | 1.0931 | 1.0958 |
| .800 | 1.0986 | 1.1014 | 1.1041 | 1.1070 | 1.1098 | 1.1127 | 1.1155 | 1.1184 | 1.1212 | 1.1241 |
| .810 | 1.1270 | 1.1299 | 1.1329 | 1.1358 | 1.1388 | 1.1417 | 1.1447 | 1.1477 | 1.1507 | 1.1538 |
| .820 | 1.1568 | 1.1599 | 1.1630 | 1.1660 | 1.1692 | 1.1723 | 1.1754 | 1.1786 | 1.1817 | 1.1849 |
| .830 | 1.1870 | 1.1913 | 1.1946 | 1.1979 | 1.2011 | 1.2044 | 1.2077 | 1.2111 | 1.2144 | 1.2178 |
| .840 | 1.2212 | 1.2246 | 1.2280 | 1.2315 | 1.2349 | 1.2384 | 1.2419 | 1.2454 | 1.2490 | 1.2526 |
| .850 | 1.2561 | 1.2598 | 1.2634 | 1.2670 | 1.2708 | 1.2744 | 1.2782 | 1.2819 | 1.2857 | 1.2895 |
| .860 | 1.2934 | 1.2972 | 1.3011 | 1.3050 | 1.3089 | 1.3129 | 1.3168 | 1.3209 | 1.3249 | 1.3290 |
| .870 | 1.3331 | 1.3372 | 1.3414 | 1.3456 | 1.3498 | 1.3540 | 1.3583 | 1.3626 | 1.3670 | 1.3714 |
| .880 | 1.3758 | 1.3802 | 1.3847 | 1.3892 | 1.3938 | 1.3984 | 1.4030 | 1.4077 | 1.4124 | 1.4171 |
| .890 | 1.4219 | 1.4268 | 1.4316 | 1.4366 | 1.4415 | 1.4465 | 1.4516 | 1.4566 | 1.4618 | 1.4670 |
| .900 | 1.4722 | 1.4775 | 1.4828 | 1.4883 | 1.4937 | 1.4992 | 1.5047 | 1.5103 | 1.5160 | 1.5217 |
| .910 | 1.5275 | 1.5334 | 1.5393 | 1.5453 | 1.5513 | 1.5574 | 1.5636 | 1.5698 | 1.5762 | 1.5825 |
| .920 | 1.5890 | 1.5956 | 1.6022 | 1.6089 | 1.6157 | 1.6226 | 1.6296 | 1.6366 | 1.6438 | 1.6510 |
| .930 | 1.6584 | 1.6659 | 1.6734 | 1.6811 | 1.6888 | 1.6967 | 1.7047 | 1.7129 | 1.7211 | 1.7295 |
| .940 | 1.7380 | 1.7467 | 1.7555 | 1.7645 | 1.7736 | 1.7828 | 1.7923 | 1.8019 | 1.8117 | 1.8216 |
| .950 | 1.8318 | 1.8421 | 1.8527 | 1.8635 | 1.8745 | 1.8857 | 1.8972 | 1.9090 | 1.9210 | 1.9333 |
| .960 | 1.9459 | 1.9588 | 1.9721 | 1.9857 | 1.9996 | 2.0140 | 2.0287 | 2.0439 | 2.0595 | 2.0756 |
| .970 | 2.0923 | 2.1095 | 2.1273 | 2.1457 | 2.1649 | 2.1847 | 2.2054 | 2.2269 | 2.2494 | 2.2729 |
| .980 | 2.2976 | 2.3223 | 2.3507 | 2.3796 | 2.4101 | 2.4426 | 2.4774 | 2.5147 | 2.5550 | 2.5988 |
| .990 | 2.6467 | 2.6996 | 2.7587 | 2.8257 | 2.9031 | 2.9945 | 3.1063 | 3.2504 | 3.4534 | 3.8002 |

| r | z |
|---|---|
| .9999 | 4.95172 |
| .99999 | 6.10303 |

Table F    Values of $\chi^2$ **Corresponding to Given Probabilities***

| Degrees of freedom ($n$) | Probability of a deviation greater than $\chi^2$ | |
|---|---|---|
| | .01 | .05 |
| 1 | 6.635 | 3.841 |
| 2 | 9.210 | 5.991 |
| 3 | 11.341 | 7.815 |
| 4 | 13.277 | 9.488 |
| 5 | 15.086 | 11.070 |
| 6 | 16.812 | 12.592 |
| 7 | 18.475 | 14.067 |
| 8 | 20.090 | 15.507 |
| 9 | 21.666 | 16.919 |
| 10 | 23.209 | 18.307 |
| 11 | 24.725 | 19.675 |
| 12 | 26.217 | 21.026 |
| 13 | 27.688 | 22.362 |
| 14 | 29.141 | 23.685 |
| 15 | 30.578 | 24.996 |
| 16 | 32.000 | 26.296 |
| 17 | 33.409 | 27.587 |
| 18 | 34.805 | 28.869 |
| 19 | 36.191 | 30.144 |
| 20 | 37.566 | 31.410 |
| 21 | 38.932 | 32.671 |
| 22 | 40.289 | 33.924 |
| 23 | 41.638 | 35.172 |
| 24 | 42.980 | 36.415 |
| 25 | 44.314 | 37.652 |
| 26 | 45.642 | 38.885 |
| 27 | 46.963 | 40.113 |
| 28 | 48.278 | 41.337 |
| 29 | 49.588 | 42.557 |
| 30 | 50.892 | 43.773 |

For larger values of ($n$), the quantity $\sqrt{2\chi^2} - \sqrt{2n-1}$ may be used as a normal deviate with unit standard deviation.

* Reprinted from Table III of R. A. Fisher: *Statistical Methods for Research Workers*, published by Oliver and Boyd Ltd., Edinburgh, by permission of the author and publishers.

# Index